Lecture Notes in Control and Information Sciences

Edited by M. Thoma and A. Wyner

For information about Vols. 1-116 please contact your bookseller or Springer-Verlag

Lecture Notes
in Control and Information Sciences 175

Editors: M. Thoma and W. Wyner

E. Rogers, D.H. Owens

Stability Analysis for Linear Repetitive Processes

Springer-Verlag
Berlin Heidelberg GmbH

Advisory Board

Authors

Eric Rogers
Advanced Systems Research Group
Dept. of Aeronautics and Astronautics
University of Southampton
Southampton, SO9 5NH
United Kingdom

David H. Owens
Centre for Systems and Control Eng.
School of Engineering
University of Exeter
Exeter, EX4 4QF
United Kingdom

ISBN 978-3-540-55264-2   ISBN 978-3-540-46999-5 (eBook)
DOI 10.1007/978-3-540-46999-5

Typesetting: Camera ready by authors

60/3020 5 4 3 2 1 0   Printed on acid-free paper

PREFACE

Repetitive, or multipass, processes are characterised by a series of sweeps, or passes, through a set of dynamics which in the simplest case is both linear and known. On each pass an output, or pass profile, is produced which acts as a forcing function on, and hence contributes to, the next pass profile. This so-called unit memory property is a special case of the more general situation where it is the previous M passes which contribute to the current one. The integer M is termed the memory length and such processes are simply termed non-unit memory. Industrial examples include long-wall coal cutting and certain metal rolling operations.

This interaction between successive pass profiles is the basic source of the unique control problem for these processes. In particular, it is possible to generate oscillations which increase in amplitude from pass to pass. Such behaviour is clearly totally unacceptable and hence appropriate control action is required.

The concept of a multipass process was first introduced in the early 1970's as a result of work at the University of Sheffield on the modelling and control of long-wall coal cutting operations. This, in turn, led to systematic attempts at controller design for these and several other industrial examples based, essentially, on appropriately modifying existing standard linear systems techniques such as Nyquist diagrams. As the number of examples increased, however, it gradually became clear that this general approach was, at best, valid only under quite restrictive conditions. Hence the need for a rigorous control theory, where stability is an obvious essential item of any such theory.

Using previously published work as a basis, this monograph presents a rigorous control theory, and associated tests, for repetitive processes with linear dynamics and a constant pass length. This is based on an abstract representation formulated in functional analysis terms by, in effect, regarding the pass profile as a point in a Banach space. All linear dynamics constant pass length examples are special cases of this abstract representation but this work concentrates on so-called differential and discrete non-unit memory linear repetitive processes which are of direct industrial relevance.

Three computationally feasible sets of stability tests are developed together with some associated properties. These then lead to some preliminary results on feedback control which are included with the general aim of stimulating further research. A central theme in the work reported here is the use of structural links with other classes of linear dynamic systems.

The work reported in this monograph was undertaken during periods when one or both of the authors were on the staff of The University of Sheffield, The Queen's University of Belfast and The University of Strathclyde. It follows on from the original work of John Edwards at Sheffield to whom we owe a great debt of gratitude as the pioneer of this area. A number of former colleagues have also made very useful suggestions, particularly Derek Collins and Ian Willson in the early days at Sheffield. Finally, we must thank Miss Yvonne Fleming for typing the final manuscript.

CONTENTS

CHAPTER 1

INTRODUCTION

The essential unique feature of a repetitive, or multipass, process can be illustrated by considering machining operations where the workpiece is processed by a series of sweeps, or passes, of a processing device. On each pass an output, or pass profile, is produced and in a repetitive process this acts as a forcing function on, and hence contributes to, the next profile. In the simplest case, therefore, the output at any point on a particular pass is a function of the independent inputs/disturbances at that point and the pass profile at the same point on the previous pass - the so-called unit memory property. Industrial examples include long-wall coal cutting and certain metal rolling operations.

Repetitive processes also exist where, in effect, the current pass profile is a function of the independent inputs/disturbances to that pass and a finite number, M, of previous pass profiles. The integer M is termed the memory length and examples in this case are simply termed 'non-unit memory'. Such examples reduce to the case described above if M is unity and hence in this sense they can be regarded as the natural generalisations of their unit memory counterparts. One major example in this case is bench mining systems.

In addition to that arising from the independent inputs and the memory property, some examples exhibit dynamic behaviour, termed interpass smoothing, between the production of successive pass profiles. One such case is long-wall coal cutting where, as a result of the basic process geometry, the machine's weight (up to 5 tonnes) causes considerable distortion to the previous pass profile as it passes over. Hence, if a physically realistic analysis is to be undertaken, a means of explicitly including this feature is clearly required.

The basic unique control problem for a repetitive process is the possible presence in the output sequence of oscillations which increase in amplitude from pass to pass. This behaviour is easily generated in simulation studies and in experimental studies on scaled models of industrial examples such as the long-wall coal cutter. Further, acceptable control of a given example clearly requires a suitable stability and control theory. This monograph describes the development of a rigorous stability theory, and associated stability tests, with particular emphasis on so-called differential and discrete non-unit memory linear repetitive processes which are of direct industrial relevance. Some highly promising results on using the developed theory and tests for feedback control of these processes is also included with the general aim of stimulating further research.

At the most general level, a repetitive process has nonlinear dynamics and a pass length which, by definition, is finite but varies from pass to pass. Hence the analysis of such a case would, at best, be a very formidable task. The special case of linear dynamics and a constant pass length is, however, tractable from an analysis standpoint and includes the vast majority of currently known examples of practical interest. Consequently this monograph will follow all published work to date and restrict attention to this special case, with the observation that progress here may also suggest approaches to other cases such as nonlinear dynamics and a constant pass length.

Early work considered only single-input/single-output systems for which an obvious intuitive approach to stability analysis and controller design is to attempt to make use of existing techniques. This basically used a single variable, termed the total distance traversed, to convert the system into an infinite length single pass process and requires following assumptions.

(i) The pass length is 'long' and hence the effects of the initial conditions on each pass can be ignored.

(ii) The effects of the previous pass dynamics can be represented by a 'long delay' term.

Other work, however, has shown that the 'resetting' action of the initial conditions on each pass can act as a form of stabilising action and hence prevent the growth of disturbances. Hence this approach is, at best, only valid 'far enough' away from either end of the pass. Further, no attempt has been made to use this approach as a basis for developing a general control policy. Instead, only the problems arising in a few well known industrial examples have been considered.

As an alternative to the, essentially classically based, approach described briefly above, suppose that rigorous stability and control theories are developed from a general base, or abstract representation, with the following core features.

(i) The effects of the initial conditions on each pass are explicitly retained.

(ii) Includes the previously studied examples as special cases but allows for others with a more complex, possibly multivariable, structure.

Then, in principle, the limitations of the classically based approach will have been removed.

To provide a suitable basis, it is obvious that any candidate abstract representation must explicitly include all of the characteristics which define these processes. This particular problem has been considered in other work which has led to the development of a suitable representation for the most general nonlinear dynamics variable pass length case. Basically, this regards the output on any pass as a point in a suitably chosen function space. Further, the sub-class of processes with linear dynamics and a constant pass length is a special case.

Using this abstract representation as a basis, a rigorous stability theory for the linear dynamics constant pass length sub-class has been developed. This consists of two distinct concepts, termed asymptotic stability and stability along the pass respectively. Further, the former is a necessary condition for the latter which is known to be required in all practical applications. The necessary and sufficient conditions for these properties are expressed in terms of the spectral radius and resolvent of the linear operator associated with the abstract representation.

In terms of applications, a number of publications have reported the results of interpreting this abstract theory for certain particular cases. For example, the results for differential and discrete non-unit memory linear repetitive processes have been documented. These, however, are not computationally feasible, where this is a common feature of the results to date for a number of other cases. Given the pivotal role of stability, the development of computationally feasible stability tests is an obvious starting point for any further control related analysis of a given case. Consequently the development of computationally feasible stability tests for differential and discrete non-unit memory linear repetitive processes forms a substantial part of the work reported in this monograph.

One approach to the analysis of repetitive systems is to exploit, where possible, structural links which may exist with other well researched classes of dynamic systems. In this work, previously documented links between differential and discrete non-unit memory linear repetitive processes and the following two classes of linear dynamic systems will form a central underlying theme.

  (i)   Standard linear systems described by the well known state-space model or transfer-function matrix.

  (ii) 2D linear (image or signal processing) systems described by the so-called Roesser state-space model.

Further, considerable use will be made of results from the stability theory of certain classes of delay differential systems. One subsidiary benefit of this general approach will be to strengthen the already known links between these areas.

The material presented in this monograph is organised into six chapters where the first of these introduces the basic features and unique control problems by reference to an industrial example. Some background results central to the analysis which follows are also developed, including a transfer-function matrix description for both the differential and discrete cases. Chapter 3 then details the stability theory based on the abstract representation and applies it to the differential and discrete processes to produce conditions suitable for the development of computationally feasible stability tests. Finally, this chapter considers the use of results from the stability analysis of systems described by the Roesser model in the same context.

Given this basis, chapter 4 develops computationally feasible stability tests for both differential and discrete processes which, in effect, use only standard linear systems tests. The end product is two systematic test procedures in each case, which are also compared from an applications standpoint with particular emphasis on Computer Aided Design aspects. This chapter also considers the use of results from the stability theory of systems described by the Roesser model and delay differential systems from the same standpoint.

Chapter 5 continues with the stability theme by developing simulation-based tests based on suitably well behaved plant step response data which is assumed to be available, or can be obtained by simulation studies. Further, it is shown that these tests produce, at no extra cost, computable information concerning the following features which are of significant importance in terms of the control of these processes.

(i) The rate of approach of the output sequence to the so-called steady, or limit, profile which is a consequence of stability.

(ii) Bounds on the performance along any pass.

This information is unique to these tests for which some initial results on extending them to processes with interpass smoothing effects are also included. These results are the first reported output on the analysis of such cases.

Following on from the previous three, chapter 6 presents the results of some initial work on controller design. In particular, three control policies are formulated from practical considerations and feedback control schemes which use either state or output information are developed. Further, some candidate design algorithms are presented together with some relevant systems theoretic properties. Finally, chapter 7 summarises progress to date and briefly outlines some possible future research topics.

CHAPTER 2

PRELIMINARIES

This chapter introduces the unique features and control problems of repetitive processes by reference to an industrial example. Some background results central to the analysis of subsequent chapters are also developed.

2.1 Unique Features and Control Problems

The essential unique feature of a repetitive, or multipass, process is the presence of a recursive action with interaction between successive outputs or pass profiles. To formalise this, first suppose, for simplicity, that the necessarily finite pass length α is constant and denote the pass profile generated over α on pass $k \geq 0$ by $Y_k(t)$, $0 \leq t \leq \alpha$. Then a repetitive process is one where $Y_k(t)$ acts as a forcing function on, and hence contributes to, the next pass profile $Y_{k+1}(t)$, $0 \leq t \leq \alpha$, $k \geq 0$. Industrial examples include long-wall coal cutting and a brief study of this case is now given to introduce the basic unique control problem for these processes.

In Great Britain, the most satisfactory, and commonly used, method of mining coal is by a process known as advanced long-wall coal cutting. Figures 2.1a and 2.1b illustrate the basic operation of the long-wall system of working in which the coal cutting machine is hauled along the entire length of the face riding on the semi-flexible structure of the armoured face conveyor, or A.F.C., which transports away the coal cut by the rotating drum. These machines generally cut in one direction only, left to right in Figures 2.1a and 2.1b, and are hauled back in reverse at high speed for the start of the new sweep, or pass, of the coal face. Between passes, the conveyor is snaked forward hydraulically, as illustrated in Figure 2.2, so that it now rests on the floor profile produced during the previous pass. During the cutting operation, the machines drum may be raised or lowered with respect to the A.F.C. by hydraulically tilting the body about a datum line on the drum or face side. The objective of this is the vertical steering of the entire long-wall installation (machine, conveyor and roof support units) to maintain it within the undulating confines of the coal seam. A nucleonic coal sensor, situated some distance behind the drum, provides the primary control signal by measuring either the floor or ceiling thickness left by the machine.

In order to obtain a simplified mathematical description of this process, consider the idealised side elevation and plan shown in Figures 2.3a and 2.3b respectively. Here the constants F,R and W represent the feet spacing, drum offset, and the width of the machine (and drum) respectively, the variable $J_{k+1}(t)$ represents the controlled drum deflection, and $e_{k+1}(t)$ denotes the height of the A.F.C. on which the machine rides. Suppose also that all angular deflections are

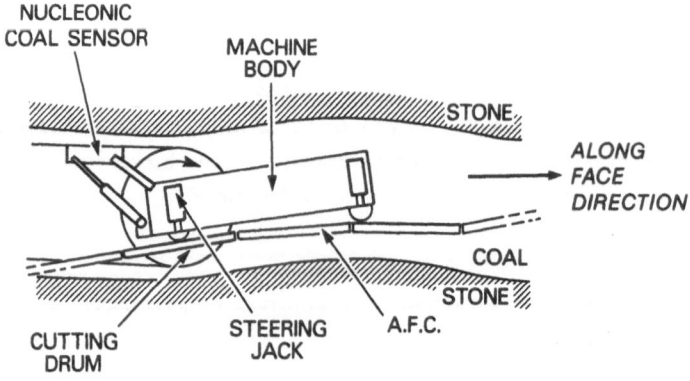

FIGURE 2.1(a)

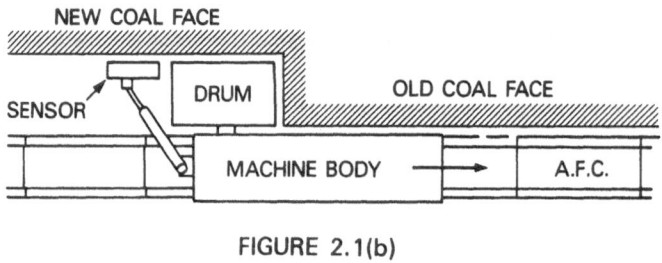

FIGURE 2.1(b)

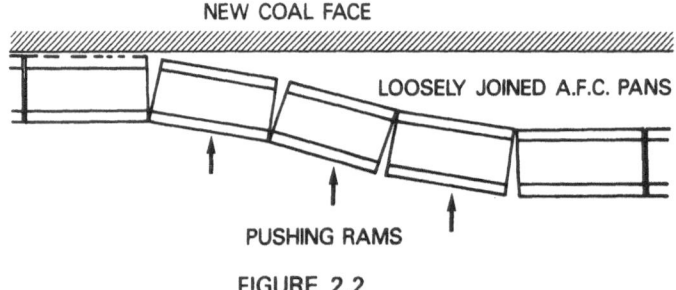

FIGURE 2.2

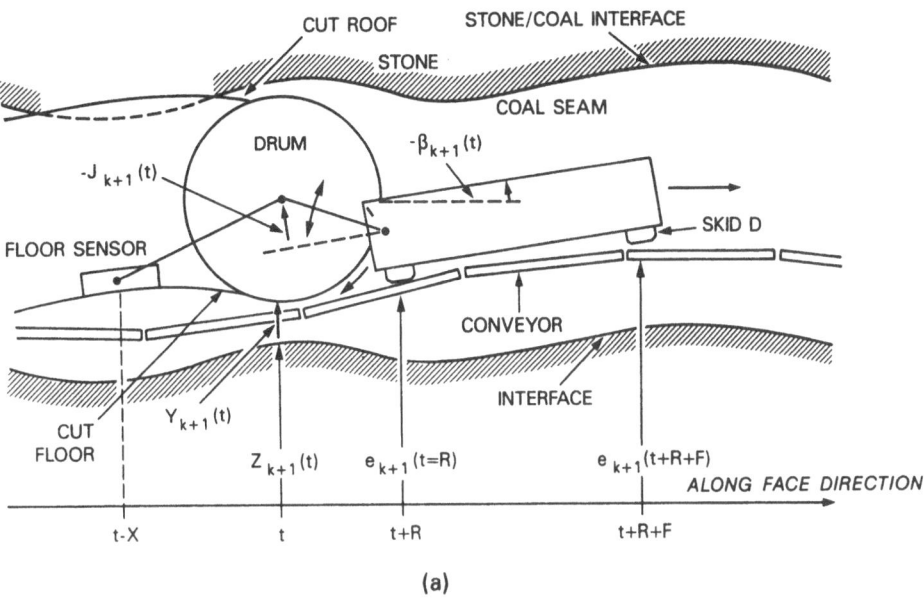

CUT ROOF STONE/COAL INTERFACE
STONE
COAL SEAM
DRUM
$-\beta_{k+1}(t)$
$-J_{k+1}(t)$
SKID D
FLOOR SENSOR
CONVEYOR
INTERFACE
CUT
FLOOR
$Y_{k+1}(t)$
$Z_{k+1}(t)$ $e_{k+1}(t=R)$ $e_{k+1}(t+R+F)$
ALONG FACE DIRECTION
t-X t t+R t+R+F

(a)

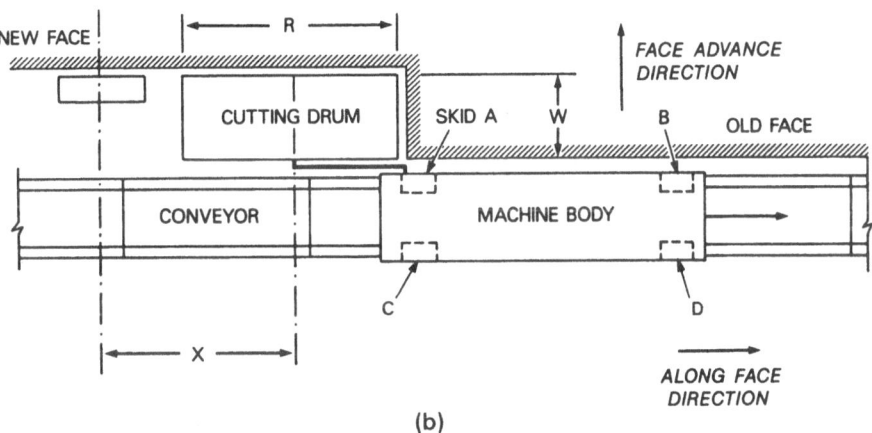

NEW FACE
R
FACE ADVANCE
DIRECTION
CUTTING DRUM SKID A W B OLD FACE
CONVEYOR MACHINE BODY
C D
X
ALONG FACE
DIRECTION

(b)

FIGURE 2.3

small. Then elementary geometrical considerations immediately yield the following description of the coal cutter dynamics

$$Y_{k+1}(t) + Z_{k+1}(t) = e_{k+1}(t + R) + W\gamma_{k+1}(t + R) + R\beta_{k+1}(t + R) + J_{k+1}(t)$$

$$0 \leq t \leq \alpha \qquad (2.1)$$

where γ, β denote the transverse and longitudinal tilts of the machine respectively and $Z_{k+1}(t)$ denotes the height of the coal/stone interface above a fixed datum plane. The transverse and longitudinal tilts of the machine are also those of the supporting conveyor structure and are given by

$$\gamma_{k+1}(t) = (e_{k+1}(t) - e_k(t))/W \qquad (2.2)$$

and

$$\beta_{k+1}(t) = (e_{k+1}(t) - e_{k+1}(t + F))/F \qquad (2.3)$$

respectively. Finally, suppose that the A.F.C. moulds itself exactly onto the cut floor upon which it rests - the so-called 'rubber conveyor' assumption. Then

$$e_{k+1}(t) = k_2(Y_k(t) + Z_k(t)) \qquad (2.4)$$

where k_2 is a positive real constant, and (2.1)-(2.4) form a complete description of the open-loop system.

Conventionally, this system is controlled by manipulation of the variable $J_{k+1}(t)$ from a delayed measurement of the floor coal thickness $Y_{k+1}(t - X)$, where X is the transport delay, or lag, equal to the distance by which the coal sensor lags behind the cutting drum. More commonly, however, the roof coal thickness is used since it can be related to $Y_{k+1}(t - X)$ on the assumption that the seam thickness is constant. Suppose also that the sensor and actuator dynamics can be neglected and a so-called fixed drum shearer is under consideration, i.e. R = 0. Then the control law in this case takes the form

$$J_{k+1}(t) = k_1(R_{k+1}(t) - Y_{k+1}(t - X)) - W\gamma_{k+1}(t) \qquad (2.5)$$

where k_1 is a positive real constant and $R_{k+1}(t)$ is a new external reference variable taken to represent the desired coal thickness on pass k + 1, k ≥ 0.

Suppose now, for simplicity, that the variable $Z_k(t)$ is set equal to zero. Then combining the above equations yields the closed-loop description

$$Y_{k+1}(t) = -k_1 Y_{k+1}(t - X) + k_2 Y_k(t) + k_1 R_{k+1}(t)$$

$$X > 0, \quad 0 \leq t \leq \alpha, \quad k \geq 0 \qquad (2.6)$$

with assumed initial conditions

$$Y_{k+1}(t) = 0, \quad -X \leq t \leq 0, \quad k \geq 0 \qquad (2.7)$$

Figure 2.4 shows the response of this closed-loop system in the special case when $k_1 = 0.8$, $k_2 = 1$, $X = 1.25$, $\alpha = 10$, to a downward unit step in $R_{k+1}(t)$ on each pass, i.e. $R_{k+1}(t) = -1$, $0 \leq t \leq 10$, $k \geq 0$. Note that the oscillations grow, or increase in amplitude, severely from pass to pass. Hence the deterioration in system performance after the first pass must be due to the fact that the cut floor profile, or dynamics, on any pass acts as a disturbance on, and hence contributes to, the dynamics of the next pass. This interaction between successive pass dynamics is the essential unique characteristic of all repetitive processes and in cases such as that of Figure 2.4 strong control action is clearly required.

Acceptable control of a repetitive process in a given case clearly requires a suitable stability and control (feedback or otherwise) theory. This monograph describes the development of a rigorous stability theory, and associated stability tests, for a special case described by a set of differential or discrete linear equations. These equations can be used to describe a number of industrial examples and in the penultimate chapter some initial results on using the developed theory and tests for feedback control of such examples will be presented.

2.2 Classical Stability Analysis - A Brief Critical Overview

If the example under consideration is single-input/single-output (SISO), an obvious intuitive approach to stability analysis and controller design is to attempt to make use of existing techniques in the form, for example, of the inverse Nyquist diagram. The essence of such an approach is to use the single variable $V = k\alpha + t$ to convert the system into an infinite length single pass process in which the relationships between variables are expressed only in terms of V. In particular, a variable, say, $Y_{k+1}(t)$, $k \geq 0$, is identified as a function of $Y(V)$ defined for $0 \leq V < +\infty$, where V is termed the total distance traversed.

Applying this approach to (2.6)-(2.7) yields

$$Y(V) = -k_1Y(V-X) + k_2Y(V-\alpha) + k_1R(V) \qquad (2.8)$$

and this repetitive process is said to be stable if, and only if, the system of (2.8) is stable in the standard sense. The repetitive process is now amenable to analysis by any of the well known classical techniques. Hence, for example, taking the Laplace transform with respect to V and making use of the inverse Nyquist diagram leads to the result that the closed-loop system is stable in the standard sense if, and only if,

$$k_1 < 1 - k_2 \qquad (2.9)$$

The above analysis can, at best, only produce useful initial guidelines since it completely ignores the considerable distortion caused to the previous pass profile by the weight (up to 5 tonnes) of the machine as it passes over. This problem is a common feature of a number of known examples of repetitive processes in that dynamic interaction, termed interpass smoothing, between passes causes

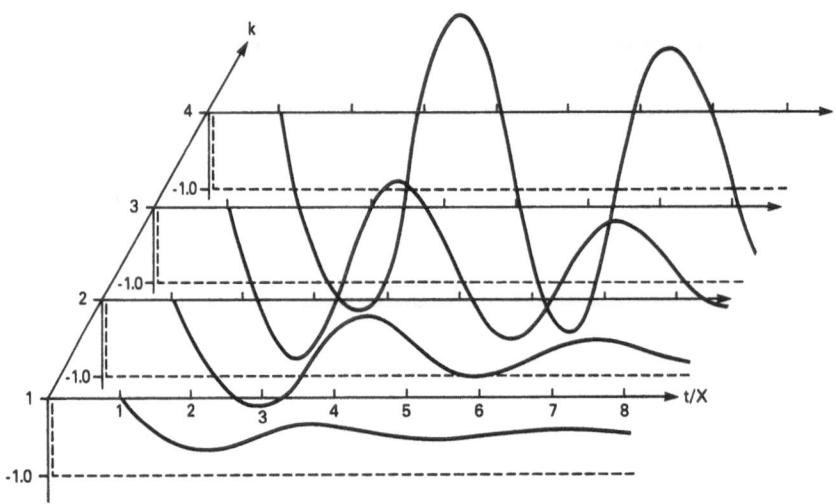

FIGURE 2.4

.distortion of the previous pass profile. It is clear, therefore, that if a physically realistic analysis of such examples is to be undertaken then a mathematical means of including this interpass smoothing is required.

In order to apply classical stability analysis and tests to repetitive processes, it is necessary to make the following assumptions.

(i) The pass length a is 'long' and hence the effects of the initial conditions on each pass can be ignored.

(ii) The effects of the previous pass dynamics can be represented by a 'long delay' term.

Intuitively, however, the 'resetting' action of the initial conditions on each pass could act as a form of stabilising action and hence prevent the growth of disturbances. In particular, it is easily shown, using a discretised form of (2.6)-(2.7) with $Y_{k+1}(t)$, $- X \leq t \leq 0$, appropriately chosen, that the initial conditions on each pass can have a crucial effect on the performance of the simplified long-wall coal cutter dynamics. Suggesting that for systems with a lag, X, on the current pass the analysis based on the concept of the total distance traversed is valid only in the range

$$ka + X \ll V \ll (k + 1)a, \ k \geq 0 \tag{2.10}$$

and for delay-free systems only in the range

$$ka \ll V \ll (k + 1)a \ , \quad k \geq 0 \tag{2.11}$$

Note also that no attempt has been made to use this approach in formulating a general control policy. Instead, attention has been restricted to the problems occuring in a few well documented industrial examples.

Summarising, therefore, the classically based approach to stability analysis and controller design discussed briefly in this section is limited by the following major factors.

(i) It completely neglects the effects of the initial conditions on each pass which are known to have a crucial effect on system stability and performance in certain cases of practical (and theoretical) interest.

(ii) No attempt has been made to develop rigorous stability and control theories for the wide range of known repetitive processes. Further, it is by no means clear that such a development is possible even for suitably well defined sub-classes.

2.3 A General Abstract Representation

As an alternative to the approach reviewed briefly in the previous section, suppose that rigorous stability and control theories are developed from a general abstract representation with the following essential features.

(i) Explicit retention of the effects of the initial conditions on each pass.

(ii) Treats the examples studied using the approach of section 2.2 as special cases and includes provision for others with a more complex, possibly multivariable, structure.

Then, in principle, the limitations of the classically based approach will have been removed.

To provide a suitable basis, it is obvious that any abstract representation must explicitly include the essential unique features. In the most general case of a variable pass length, these can be summarised as follows and are also illustrated in Figure 2.5.

(a) A number of passes through a known set of dynamics.

(b) Each pass is characterised by a pass length , α_k, which may vary from pass to pass and a pass profile $Y_k(t)$ defined on $0 \leq t \leq \alpha_k$. Note that the pass profile need not be a scalar quantity.

(c) An initial pass profile $Y_0(t)$ defined on $0 \leq t \leq \alpha_0$, where α_0 is the initial pass length. The function Y_0 plays the role of an initial condition for the process.

(d) Each pass will be subject to its own boundary conditions, disturbances and control inputs.

(e) The process is unit memory, i.e. the dynamics on pass $k + 1$ depend only on the independent inputs to that pass and the pass profile on the previous pass k.

Given (a)-(e), suppose that $Y_k(t)$ is regarded as point in a suitably chosen function space. In particular, suppose that

$$Y_k \in E_{\alpha_k} \quad , \quad k \geq 0 \qquad\qquad (2.12)$$

where E_{α_k} denotes a Banach space. Then a general abstract model of these processes can be formulated as a recursion relation of the form

$$Y_{k+1} = f_{k+1}(Y_k) \quad , \quad k \geq 0 \qquad\qquad (2.13)$$

(where f_{k+1} is an abstract mapping of E_{α_k} into $E_{\alpha_{k+1}}$) together with a rule for updating the pass length α_k of the form

$$\alpha_{k+1} = g_{k+1}(\alpha_k, Y_k, Y_{k+1}), \quad k \geq 0 \qquad\qquad (2.14)$$

Repetitive processes also exist, for example so-called bench mining systems, where the current pass profile is a function of the independent inputs to that pass and a finite number, $M > 1$, of previous pass profiles. The integer M is termed the memory length and such processes are designated as 'non-unit memory of length M' or, more simply, 'non-unit memory'. Such processes are easily accommodated within the general structure of (2.13)-(2.14). Formally, all that is required is to replace these equations by

$$Y_{k+1} = f_{k+1}(Y_k, Y_{k-1}, \ldots, Y_{k+1-M}), \quad k \geq 0 \qquad\qquad (2.15)$$

and

$$\alpha_{k+1} = g_{k+1}(\alpha_k, \alpha_{k-1}, \ldots, \alpha_{k+1-M}, Y_{k+1}, \ldots, Y_{k+1-M}), \quad k \geq 0 \qquad\qquad (2.16)$$

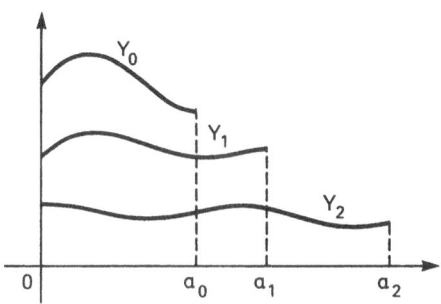

FIGURE 2.5

respectively. This formulation is unnecessary, however, if the ordered set $(Y_k, Y_{k-1}, \ldots, Y_{k+1-M})$ is regarded as a 'pass profile' in the product space $E_{\alpha_k} \times E_{\alpha_{k-1}} \times \ldots \times E_{\alpha_{k+1-M}}$, i.e.

$$(Y_k, Y_{k-1}, \ldots, Y_{k+1-M}) \in E_{\alpha_k} \times E_{\alpha_{k-1}} \times \ldots \times E_{\alpha_{k+1-M}} \tag{2.17}$$

In which case (2.15) and (2.16) become

$$(Y_{k+1}, Y_k, \ldots, Y_{k+2-M}) = (f_{k+1}(Y_k, \ldots, Y_{k+1-M}), Y_k, \ldots, Y_{k+2-M}) \tag{2.18}$$

and

$$\alpha_{k+1} = g_{k+1}(\alpha_k, \ldots, \alpha_{k+1-M}, Y_{k+1}, \ldots, Y_{k+1-M}) \tag{2.19}$$

respectively which have an identical structure to (2.15) and (2.16). Now, however, M points $Y_0, Y_{-1}, \ldots, Y_{1-M}$ are required to define the initial profile.

Any analysis of the abstract model defined above would clearly be a formidable task. A difficulty which can be avoided by noting that the vast majority of processes of practical interest are linear and of constant pass length. Hence from this point onwards attention will be restricted to linear processes with

$$\alpha_k = \alpha, \quad k \geq 0 \tag{2.20}$$

The following general definition characterises the unit memory version in this case.

<u>Definition 2.3.1:</u> A linear repetitive process $S(E_\alpha, W_\alpha, L_\alpha)$ of constant pass length $\alpha > 0$ consists of a Banach space E_α, a linear subspace W_α of E_α, and a bounded linear operator L_α of E_α into itself. The system dynamics are described by linear recursion relations of the form

$$Y_{k+1} = L_\alpha Y_k + b_{k+1}, \quad k \geq 0 \tag{2.21}$$

where $Y_k \in E_\alpha$ is the pass profile on pass k and $b_{k+1} \in W_\alpha$, $k \geq 0$. Here the term $L_\alpha Y_k$ represents the contribution from pass k to pass k + 1 and b_{k+1} represents initial conditions, disturbances and control input effects. ∎

In the non-unit memory case, let L_α^j, $1 \leq j \leq M$, be bounded linear operators mapping E_α into itself. Then the most general representation of a constant pass length non-unit memory linear repetitive process with memory length M takes the form

$$Y_{k+1} = L_\alpha^1 Y_k + L_\alpha^2 Y_{k-1} + \ldots + L_\alpha^M Y_{k+1-M} + b_{k+1} \tag{2.22}$$

where $Y_k \in E_\alpha$, $k \geq 1 - M$, $b_{k+1} \in W_\alpha \subset E_\alpha$. Note also that (2.22) reduces to (2.21) with $L_\alpha \equiv L_\alpha^1$ if M = 1 and hence it can be regarded as the natural non-unit memory generalisation. Further, using (2.18), it can be regarded as a process of the form (2.21) in the product space $E_\alpha^M = E_\alpha \times E_\alpha \times \ldots \times E_\alpha$ (M times) by writing it in the 'companion form'

$$
\begin{bmatrix} Y_{k+2-M} \\ \\ \\ Y_{k+1} \end{bmatrix} = \begin{bmatrix} 0 & I & & & 0 \\ & & & & \\ 0 & & & 0 & I \\ & & & & \\ L_\alpha^M & L_\alpha^{M-1} & & L_\alpha^2 & L_\alpha^1 \end{bmatrix} \begin{bmatrix} Y_{k+1-M} \\ \\ \\ Y_k \end{bmatrix} + \begin{bmatrix} 0 \\ \\ 0 \\ \\ b_{k+1} \end{bmatrix}, \quad k \geq 0
$$

(2.23)

and using the notation

$$
L_\alpha = \begin{bmatrix} 0 & I & & 0 \\ 0 & & 0 & I \\ L_\alpha^M & L_\alpha^{M-1} & L_\alpha^2 & L_\alpha^1 \end{bmatrix}
$$

(2.24)

Hence results derived for the unit memory case can immediately be applied to the non-unit memory generalisation.

To illustrate the generality of (2.23)-(2.24) (and (2.21)), the following examples are now considered.

Example 2.3.1 - A delay - algebraic system - The scalar equation

$$
Y_{k+1}(t) = - k_0 Y_{k+1}(t - X) + k_1 Y_k(t) + k_0 R_{k+1}(t)
$$

$$
0 \leq t \leq \alpha, \quad k \geq 0
$$

$$
Y_{k+1}(t) = 0, \quad - X \leq t \leq 0
$$

(2.25)

where k_0 and k_1 are constants, has been shown to represent physical examples of repetitive processes such as long-wall coal cutting and metal rolling. This equation has the structure of a unit memory linear repetitive process of pass length α, with $E_\alpha = W_\alpha$ the vector space of continuous functions on $[0,\alpha]$ satisfying the initial condition $Y(0) = 0$ and norm

$$
||Y|| = \max_{0 \leq t \leq \alpha} |Y(t)|
$$

(2.26)

The operator L_α is defined by expressing $Y_1 = L_\alpha Y_0$ in the form

$$
Y_1(t) = - k_0 Y_1(t - X) + k_1 Y_0(t), \quad 0 \leq t \leq \alpha
$$

$$
Y_1(t) = 0, \quad - X \leq t \leq 0
$$

(2.27)

Example 2.3.2 - Matrix recursion relations - The discrete state vector model

$$
X_{k+1} = AX_k + BU_k, \quad X_k \in R^n, \quad U_k \in R^\ell, \quad k \geq 0
$$

(2.28)

can be regarded as a unit memory linear repetitive process with $E_\alpha = R^n$, $W_\alpha = $ range of B and $b_{k+1} = BU_k$, $k \geq 0$.

Example 2.3.3 - A differential non-unit memory linear repetitive process - The state-space model in this case has the form

$$\dot{X}_{k+1}(t) = AX_{k+1}(t) + BU_{k+1}(t) + \sum_{j=1}^{M} B_{j-1}Y_{k+1-j}(t)$$

$$Y_{k+1}(t) = CX_{k+1}(t) + D_0U_{k+1}(t) + \sum_{j=1}^{M} D_j Y_{k+1-j}(t)$$

$$X_{k+1}(t) \in R^n, \ Y_{k+1}(t) \in R^m, \ U_{k+1}(t) \in R^\ell$$

$$0 \le t \le \alpha, \ X_{k+1}(0) = d_{k+1}, \ k \ge 0 \qquad (2.29)$$

To write (2.29) in the form $S(E_\alpha, W_\alpha, L_\alpha)$, as defined by (2.23)-(2.24), first note that

$$Y_{k+1}(t) = C\int_0^t e^{A(t-\tau)}\{\sum_{j=1}^{M} B_{j-1}Y_{k+1-j}(\tau) + BU_{k+1}(\tau)\}d\tau$$

$$+ C \, e^{At}d_{k+1} + D_0U_{k+1}(t) + \sum_{j=1}^{M} D_j Y_{k+1-j}(t), \ 0 \le t \le \alpha, \ k \ge 0$$

$$(2.30)$$

Further, consider the problem in the context of the Banach space $E_\alpha = C_m(0,\alpha)$ of bounded continuous mappings of the interval $0 \le t \le \alpha$ into the vector space of real m-vectors R^m with norm

$$||Y|| = \sup_{0 \le t \le \alpha} ||Y(t)||_m \qquad (2.31)$$

where $||.||_m$ is any convenient norm in R^m, e.g. $||P||_m = \max_{1 \le i \le m} |P_i|$. Then L_α^j, $1 \le j \le M$, is defined by the relation

$$(L_\alpha^j Y)(t) = C\int_0^t e^{A(t-\tau)} B_{j-1}Y(\tau)d\tau + D_j Y(t), \ 0 \le t \le \alpha \qquad (2.32)$$

and b_{k+1} by

$$b_{k+1} = C\int_0^t e^{A(t-\tau)}BU_{k+1}(\tau)d\tau + D_0U_{k+1}(t) + C \, e^{At}d_{k+1}, \ 0 \le t \le \alpha \qquad (2.33)$$

Finally, if the system initial conditions d_{k+1}, $k \ge 0$, of interest lie in a subspace, W, of R^n and the control inputs $U_{k+1}(t)$, $k \ge 0$, are assumed to be piecewise continuous, $W_\alpha \subset E_\alpha$ can be obtained by evaluating (2.33) for all such d_{k+1} and U_{k+1}.

Example 2.3.4 - A differential unit memory linear repetitive process - Set M = 1 in the analysis of example 2.3.3 to obtain this special case.

Example 2.3.5 - A differential unit memory linear repetitive process with interpass smoothing - Consider, for simplicity, the unit memory case and hence M = 1 in (2.29) of example 2.3.3. Then one possible method of modelling the effects of interpass smoothing on the process dynamics is to assume that the pass profile at any point t on pass k + 1 is a function of the state and inputs at this point on pass k + 1 and of the complete pass profile on pass k. For example, a candidate representation is

$$\dot{X}_{k+1}(t) = AX_{k+1}(t) + BU_{k+1}(t) + B_0 \int_0^\alpha K(t,\tau)Y_k(\tau)d\tau$$

$$Y_{k+1}(t) = CX_{k+1}(t)$$

$$0 \le t \le \alpha, \quad X_{k+1}(0) = d_{k+1}, \quad k \ge 0 \tag{2.34}$$

where the interpass interaction term $B_0 \int_0^\alpha K(t,\tau)Y_k(\tau)d\tau$ represents a 'smoothing out' of the previous pass profile in a manner governed by the properties of the kernel $K(t,\tau)$. Note that the particular choice of

$$K(t,\tau) = \delta(t - \tau)I_m \tag{2.35}$$

where δ denotes the Dirac delta function reduces (2.34) to the case of example 2.3.4. It is now easily verified that (2.34) is a linear repetitive process in $E_\alpha = C_m(0,\alpha)$ with

$$(L_\alpha Y)(t) = C\int_0^t e^{A(t-\tau)} B_0 \int_0^\alpha K(\tau,t')Y(t')dt'd\tau , \quad 0 \le t \le \alpha \tag{2.36}$$

and

$$b_{k+1} = C\int_0^t e^{A(t-t')}BU_{k+1}(t')dt' + C e^{At}d_{k+1}, \quad 0 \le t \le \alpha \tag{2.37}$$

This approach can also be used to study the effects of interpass smoothing on the dynamic behaviour of processes described by the equation of example 2.3.1.

The next example provides a link between the unit memory version of example 2.3.3 and standard linear systems with a delay in the state.

Example 2.3.6 - A differential unit memory linear repetitive process with interaction between pass profiles and pass boundary conditions - Set $M = 1$, $n = m$, $D_0 = 0$, $D_1 = 0$, $C = I_n$ in (2.29) and consider the case when $X_{k+1}(0) = d_{k+1}$, $k \ge 0$, is replaced by pass dependent initial conditions of the form

$$X_{k+1}(0) = d_{k+1} + K_0 X_k(0) + \sum_{j=1}^q K_j X_k(t_j) + \int_0^\alpha K(t)X_k(t)dt \tag{2.38}$$

where $K_0, K_1, \ldots, K_q$ are constant $n \times n$ matrices, $K(t)$ is a piecewise continuous $n \times n$ matrix function of t on $0 \le t \le \alpha$ and $0 \le t_1 \le t_2 \le \ldots \le t_q \le \alpha$ are q sample points. Then this process is a linear repetitive process in $E_\alpha = C_n(0,\alpha)$ since it is easily verified that the unit memory version of the construction given in example 2.3.3 still holds in this case with L_α defined by

$$(L_\alpha Y)(t) = \int_0^t e^{A(t-t')}B_0 Y(t')dt' + e^{At}\hat{Y}, \quad 0 \le t \le \alpha \tag{2.39}$$

where

$$\hat{Y} = K_0 Y(0) + \sum_{j=1}^q K_j Y(t_j) + \int_0^\alpha K(t)Y(t)dt \tag{2.40}$$

A class of delay differential systems in R^n can be modelled by the state-space equations

$$\dot{X}(t) = AX(t) + B_0 X(t-\alpha) + BU(t) , \quad t \geq 0$$

$$X(t-\alpha) := X_0(t) , \quad 0 \leq t \leq \alpha \qquad (2.41)$$

where A, B_0, B are constant $n \times n$, $n \times n$ and $n \times \ell$ matrices respectively. If the delay α is interpreted as a pass length then it is obvious that these systems have certain structural similarities to linear repetitive processes described by a set of recursive differential equations. In particular, introduce the change of variables

$$U_{k+1}(t) = U(k\alpha + t) \qquad (2.42)$$

$$X_k(t) = X((k-1)\alpha + t), \quad 0 \leq t \leq \alpha, \quad k \geq 0 \qquad (2.43)$$

and define the pass profiles as $Y_k = X_k$, $k \geq 0$. Then (2.41) can be written as a repetitive process of the form defined by example 2.3.6 with boundary conditions

$$X_{k+1}(0) = X_k(\alpha), \quad k \geq 0 \qquad (2.44)$$

i.e. a special case of (2.38) with $q = 1$, $K_1 = I_n$, $t_1 = \alpha$, $d_{k+1} = 0$, $K_0 = 0$ and $K(t) \equiv 0$.

The next two examples are the natural discrete analogues of the processes defined in examples 2.3.3 and 2.3.4.

Example 2.3.7 - A discrete non-unit memory linear repetitive process - This is the natural discrete analogue of the process of example 2.3.3 and has state-space model

$$X_{k+1}(P + 1) = \Phi X_{k+1}(P) + \Delta U_{k+1}(P) + \sum_{j=1}^{M} \Delta_{j-1} Y_{k+1-j}(P)$$

$$Y_{k+1}(P) = CX_{k+1}(P) + D_0 U_{k+1}(P) + \sum_{j=1}^{M} D_j Y_{k+1-j}(P)$$

$$X_{k+1}(P) \in R^n, \quad Y_{k+1}(P) \in R^m, \quad U_{k+1}(P) \in R^\ell$$

$$0 \leq P \leq \alpha, \quad X_{k+1}(0) = d_{k+1}, \quad k \geq 0 \qquad (2.45)$$

Further, define the pass profile on pass k to be the ordered set

$$Y_k = \{Y_k(0), Y_k(1), \ldots, Y_k(\alpha)\} \qquad (2.46)$$

and regard it as a point in the product space $E_\alpha = R^m \times R^m \times \ldots \times R^m$ with norm

$$||Y_k|| = \max_{0 \leq P \leq \alpha} ||Y_k(P)||_m \qquad (2.47)$$

where, as in example 2.3.3, $||.||_m$ is any convenient norm in R^m. Then it is easily shown that this process can be written in the form $S(E_\alpha, W_\alpha, L_\alpha)$, as defined by (2.23) - (2.24), with L_α^j, $1 \leq j \leq M$, defined by

$$(L_\alpha^j Y)(P) = \sum_{r=0}^{P-1} C\Phi^{P-1-r} \Delta_{j-1} Y(r) + D_j Y(P), \quad 0 \leq P \leq \alpha \qquad (2.48)$$

and the disturbance b_{k+1} by

$$b_{k+1} = \sum_{r=0}^{P-1} C\Phi^{P-1-r}\Delta U_{k+1}(r) + D_0\,U_{k+1}(P) + C\Phi^P d_{k+1},\ 0 \le P \le \alpha \qquad (2.49)$$

Example 2.3.8 - A discrete unit memory linear repetitive process - Set
M = 1 in the analysis of example 2.3.7 to obtain this special case.

The situations covered by examples 2.3.5 and 2.3.6 also extend in a natural
manner to the discrete case and hence the details are omitted. Further, examples
2.3.1, 2.3.3, 2.3.4, 2.3.7 and 2.3.8 have direct industrial relevance in the
modelling for initial simulation and control studies of industrial examples such as
long-wall coal cutting, metal rolling and bench mining systems. Hence the examples
used in the remainder of this work will be exclusively drawn from these and/or
extensions to include, for example, interpass smoothing effects.

2.4 Structural Links with other Dynamic Systems

One approach to the analysis of repetitive systems is to exploit, where
possible, structural links which may exist with other well researched classes of
dynamic systems. In this work such links between the processes of examples 2.3.3 -
2.3.4 and 2.3.7 - 2.3.8 and two other classes of linear dynamic systems will be
extensively used. To introduce the first of these, consider the differential
non-unit memory linear repetitive process of example 2.3.3 and suppose that the
following operations are applied to its state-space model:
(i) The previous pass terms are deleted or, equivalently, $B_{j-1} = 0$,

   $D_j = 0$, $1 \le j \le M$.
(ii) The subscript k + 1 is dropped.
(iii)The concept of a pass length is irrelevant.
Then (2.29) reduces to

$$\dot{X}(t) = AX(t) + BU(t)$$
$$Y(t) = CX(t) + D_0 U(t)$$
$$X(0) = d \qquad (2.50)$$

which is just the well known state-space model from standard, or conventional,
linear systems theory. Within the repetitive systems framework, (2.50) is termed
the derived conventional linear system and, for notational simplicity, will be
denoted by $L_D(A,B,C,D_0)$ from this point onwards. Use will also be made of its
transfer-function matrix description in the case of d = 0

$$Y(s) = G_0(s)U(s) \qquad (2.51)$$

with

$$G_0(s) = C(sI_n - A)^{-1}B + D_0 \qquad (2.52)$$

In addition to $L_D(A,B,C,D_0)$, use will also be made of the so-called associated
conventional linear systems of (2.29) defined as

$$X(t) = AX(t) + B_{j-1}Y^{1-j}(t)$$
$$W^j(t) = CX(t) + D_jY^{1-j}(t)$$
$$X(0) = 0, \quad 1 \leq j \leq M \tag{2.53}$$

Suppose also that $d_{k+1} = 0$, $k \geq 0$. In which case the ith element of (2.53) has, in effect, been obtained from (2.29) by setting $B = 0$, $D_0 = 0$, $B_{j-1} = 0$, $D_j = 0$, $1 \leq j \neq i \leq M$, ignoring the pass length α, and dropping the pass subscript. Equivalently, (2.53) can be regarded as describing the contribution of pass profile $k + 1 - j$ to the current one. To see this, restrict t to $[0,\alpha]$ and set $Y^{1-j}(t)$ equal to pass profile $k + 1 - j$. For notational convenience, the jth, $1 \leq j \leq M$, element of (2.53) will be denoted by $L_A^j(A, B_{j-1}, C, D_j]$ from this point onwards and use will also be made of the corresponding transfer-function matrix description

$$W^j(s) = G_j(s)Y^{1-j}(s) \tag{2.54}$$

with

$$G_j(s) = C(sI_n - A)^{-1}B_{j-1} + D_j \tag{2.55}$$

Finally, in the compact notation, the derived and associated conventional linear systems for the discrete non-unit memory linear repetitive process of example 2.3.7 are defined by $L_D(\Phi,\Delta,C,D_0)$ and $L_A^j(\Phi,\Delta_{j-1},C,D_j]$, $1 \leq j \leq M$, respectively with corresponding transfer-function matrices

$$G_0(z_1) = C(z_1I_n - \Phi)^{-1}\Delta + D_0 \tag{2.56}$$

and

$$G_j(z_1) = C(z_1I_n - \Phi)^{-1}\Delta_{j-1} + D_j \tag{2.57}$$

The second area from which structural links are exploited is that of 2D linear systems described by the so-called Roesser state-space model. This has the following form for systems recursive in the positive quadrant

$$X_h(i + i,j) = A_1X_h(i,j) + A_2X_v(i,j) + B_1U(i,j)$$
$$X_v(i,j + 1) = A_3X_h(i,j) + A_4X_v(i,j) + B_2U(i,j)$$
$$Y(i,j) = C_1X_h(i,j) + C_2X_v(i,j) + DU(i,j) \tag{2.58}$$

Here i,j are positive integer valued horizontal and vertical coordinates, $X_h \in R^{n_1}$, $X_v \in R^{n_2}$ are vectors which propagate information in the horizontal and vertical directions respectively, $U \in R^\ell$ and $Y \in R^m$ are vector inputs and outputs respectively and $A_1, A_2, A_3, A_4, B_1, B_2, C_1, C_2$ and D are real constant matrices of appropriate dimensions. Systems described by this state-space model have been extensively studied in recent years using both state-space and transfer-function matrix techniques where the latter is two variable.

Comparing (2.58) with the repetitive processes described in section 2.3, and in particular the discrete process of example 2.3.8, indicates that, despite notational differences, these repetitive processes have clear structural similarities with 2D systems described by the Roesser model. In particular, the model of example 2.3.8 is a Roesser model where

(i)   X, the current pass state vector, plays the role of horizontally transmitted information;

(ii) Y, the current pass output vector, plays the role of vertically transmitted information; and

(iii) the final equation in (2.58) is redundant but could, if required, be used to represent other algebraic measurement equations associated with the application under consideration.

This similarity can be further highlighted by considering Figure 2.6 where the vertical axis is taken to represent the evolution of a discrete process from pass to pass and the horizontal axis is taken to represent the evolution of the process state along the pass. In particular, note that each point in the (k,P) plane of this figure is associated with a state $X_k(P)$, an output $Y_k(P)$ and a control input $U_k(P)$. Hence Figure 2.6 illustrates the evolution in both k and P of the state variables in terms of their values, the values of the outputs at points (j,P), j = k - 1, and the control input. This evolution of the 2D/repetitive process is uniquely specified by the boundary conditions along the axes $\{(k,0): k \geq 1\}$ and $\{(0,P): 0 \leq P \leq a\}$, i.e. the state initial conditions on each pass and the initial pass profile.

2.5   Transfer-Function Matrix Description

The corresponding transfer-function matrices play a central role in the analysis and control of conventional, or standard, linear systems and 2D linear systems described by the Roesser state-space model. Hence, given the discussion of section 2.4, it is to be expected that a similar role exists for appropriately defined transfer-function matrices in the analysis and control of the repetitive processes of examples 2.3.3 and 2.3.7. Before proceeding to consider this matter further, a number of important preliminary results and observations are required, the first of which is the fact that these processes are 'well posed' in the sense that they map sequences of inputs into sequences of outputs and each has a solution which is unique. Secondly, they exhibit multipass causality. In particular, noting that the following extends in a natural manner to example 2.3.7, consider the differential process of example 2.3.3. Then in this case multipass causality means that the output, $Y_k(t)$, at any time t on pass k does not depend on information from the following sets, see also Figure 2.7,

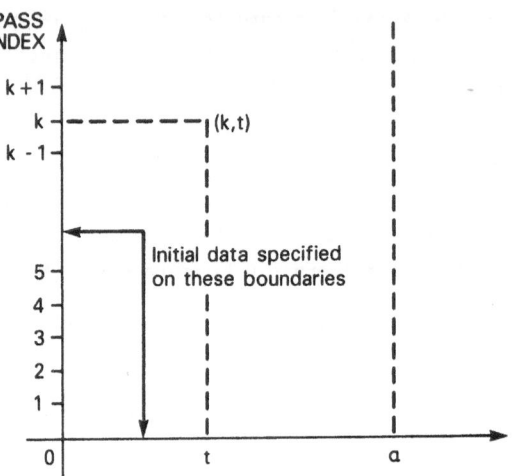

FIGURE 2.6

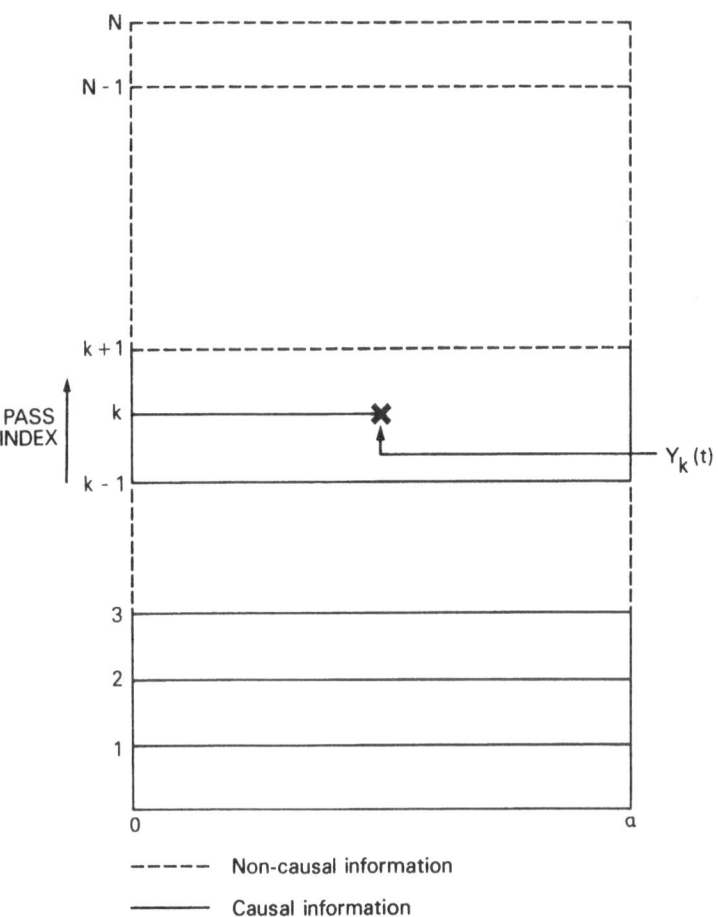

FIGURE 2.7

$$X = \{X_k(\tau): \ t < \tau \leq a\} \cup \{X_\ell(t): \ 0 \leq t \leq a, \ \ell > k\}$$

$$D = \{d_\ell: \ \ell > k\}$$

$$U = \{U_k(\tau): \ t < \tau \leq a\} \cup \{U_\ell(t): \ 0 \leq t \leq a, \ \ell > k\}$$

$$Y = \{Y_k(\tau): \ t < \tau \leq a\} \cup \{Y_\ell(t): \ 0 \leq t \leq a, \ \ell > k\} \tag{2.59}$$

Continuing with the differential case, note that two parameters are required to specify a variable in (2.29), i.e. the distance, or time, t, along a particular pass and the pass number k. Hence any transfer-function matrix description must be two variable, or 2D, in structure. Consider also the interpass dependence and regard the pass length a as a 'sample period' or interval. Then the following definition can be regarded as the natural repetitive process generalisation of the well known z-transform from discrete conventional linear systems theory.

<u>Definition 2.5.1:</u> The 'z-transforms' of the sequences $U_{k+1}(t)$,

$X_{k+1}(t)$ and $Y_{k+1}(t)$, $0 \leq t \leq a$, $k \geq 0$ are defined by

$$U(t,z) = U_1(t) + z^{-1}U_2(t) + z^{-2}U_3(t) + \ldots \tag{2.60}$$

$$X(t,z) = X_1(t) + z^{-1}X_2(t) + z^{-2}X_3(t) + \ldots \tag{2.61}$$

and

$$Y(t,z) = Y_1(t) + z^{-1}Y_2(t) + z^{-2}Y_3(t) + \ldots \tag{2.62}$$

respectively. ∎

Results on the convergence and existence properties of (2.60)-(2.62) are contained in the following result.

<u>Lemma 2.5.1:</u> Suppose that the terms in (2.60)-(2.62) are bounded in the sense that there exists real numbers $M_i > 0$, $\lambda_i > 0$, $1 \leq i \leq 3$, such that

$$||U_k(.)|| \leq M_1 \, \lambda_1^{k-1}, \quad k \geq 1 \tag{2.63}$$

$$||X_k(.)|| \leq M_2 \, \lambda_2^{k-1}, \quad k \geq 1 \tag{2.64}$$

and

$$||Y_k(.)|| \leq M_3 \, \lambda_3^{k-1}, \quad k \geq 1 \tag{2.65}$$

where $||.||$ is chosen as any suitable norm in E_a. Then (2.60)-(2.62) converge absolutely in the regions $|z| > \lambda_1$, $|z| > \lambda_2$ and $|z| > \lambda_3$ respectively.

<u>Proof:</u> Consider (2.60) and take the norm to yield

$$||U(.,z)|| \leq ||U_1(.)|| + |z^{-1}| \ ||U_2(.)|| + \ldots$$

$$\leq M_1 (1 + \frac{\lambda_1}{|z|} + (\frac{\lambda_1}{|z|})^2 + \ldots) \tag{2.66}$$

and hence absolute convergence since

$$||U(.,z)|| \leq \frac{M_1}{(1 - \frac{\lambda_1}{|z|})} < + \infty \tag{2.67}$$

provided $|z| > \lambda_1$. The proofs for (2.61) and (2.62) follow in a similar manner and are hence omitted. ∎

At this stage, define $\frac{\partial}{\partial t} X(t,z)$ as

$$\frac{\partial}{\partial t} X(t,z) = \frac{\partial}{\partial t} X_1(t) + z^{-1} \frac{\partial}{\partial t} X_2(t) + z^{-2} \frac{\partial}{\partial t} X_3(t) + \ldots \tag{2.68}$$

and consider, without loss of generality, the special case of zero initial pass profiles and zero state initial conditions on each pass, i.e.

$$Y_{1-j}(t) = 0, \quad 0 \leq t \leq \alpha, \ 1 \leq j \leq M \tag{2.69}$$

$$d_{k+1} = 0, \quad k \geq 0 \tag{2.70}$$

Hence $X(0,z) = 0$ and the 'z-transform' of (2.29) in this case is easily shown to be

$$\frac{\partial}{\partial t} X(t,z) = (A + B(z)(I_m - D(z))^{-1}C) X(t,z) + \{B +$$

$$B(z)(I_m - D(z))^{-1}D_0\} U(t,z) \tag{2.71}$$

$$Y(t,z) = (I_m - D(z))^{-1} \{CX(t,z) + D_0 U(t,z)\} \tag{2.72}$$

where

$$B(z) = \sum_{j=1}^{M} B_{j-1} z^{-j}, \quad D(z) = \sum_{j=1}^{M} D_j z^{-j} \tag{2.73}$$

and the term $(I_m - D(z))$ is always invertible since $\lim_{|z| \to +\infty}$ $(I_m - D(z)) = I_m$ which is obviously nonsingular. Note also that this result is obvious if z^{-1} is regarded as a backward shift operator.

Given (2.71)-(2.73), consider the problem of using the 'z-transform' to solve for a sequence of pass profiles in the presence of a known input sequence $U_{k+1}(t)$, $0 \leq t \leq \alpha$, $k \geq 0$. In which case it follows immediately that this can be achieved by solving (2.71) for $X(t,z)$, substituting the result in (2.72) to obtain $Y(t,z)$ and then expanding the result as a power series to obtain the pass profiles in the order $\{Y_1, Y_2, Y_3, \ldots\}$.

One method of solving for $X(t,z)$, and hence $Y(t,z)$, in (2.71)-(2.73) would be to employ the Laplace transform. Note, however, that the variables $U_j(t)$, $X_j(t)$ and $Y_j(t)$, $j \geq 1$, of the series $U(t,z)$, $X(t,z)$ and $Y(t,z)$ respectively are only defined on the finite interval $[0,\alpha]$ but use of the Laplace transform would require that they be defined on $[0,+\infty)$. Hence it would appear that the Laplace transform cannot be used in this particular situation. The fact that this is not the case is a direct result of the multipass causality of (2.29) as defined by (2.59). In particular, by multipass causality, the result will be unaffected if the Laplace transform is applied to arbitrary extensions of the variables listed above from

$[0,\alpha]$ to $[0,+\infty)$, provided, of course, that these extensions satisfy the necessary existence conditions.

Suppose, therefore, that the variables $U_j(t)$, $X_j(t)$ and $Y_j(t)$, $j \geq 1$, have been suitably extended from $[0,\alpha]$ to $[0,+\infty)$ and let the same symbols denote these extensions. Then the Laplace transforms, or 's transforms', are defined as follows.

<u>Definition 2.5.2:</u> The 's transforms' of the series $U(t,z)$, $X(t,z)$ and $Y(t,z)$ are defined by

$$U(s,z) = \mathscr{L}U(t,z) = \mathscr{L}U_1(t) + z^{-1} \mathscr{L}U_2(t) + z^{-2} \mathscr{L}U_3(t) + \ldots \tag{2.74}$$

$$X(s,z) = \mathscr{L}X(t,z) = \mathscr{L}X_1(t) + z^{-1} \mathscr{L}X_2(t) + z^{-2} \mathscr{L}X_3(t) + \ldots \tag{2.75}$$

and

$$Y(s,z) = \mathscr{L}Y(t,z) = \mathscr{L}Y_1(t) + z^{-1} \mathscr{L}Y_2(t) + z^{-2} \mathscr{L}Y_3(t) + \ldots \tag{2.76}$$

respectively where $\mathscr{L}$ denotes the Laplace transform with respect to the along the pass variable t. ∎

Results on the convergence and existence properties of (2.74)-(2.76) are contained in the following result.

<u>Lemma 2.5.2:</u> Suppose that there exists real numbers $M_i > 0$, $\beta_i > 0$ and $\lambda_i > 0$, $1 \leq i \leq 3$, such that

$$||U_j(t)||_p \leq M_1 e^{\beta_1 t} \lambda_1^{j-1} \tag{2.77}$$

$$||X_j(t)||_p \leq M_2 e^{\beta_2 t} \lambda_2^{j-1} \tag{2.78}$$

and

$$||Y_j(t)||_p \leq M_3 e^{\beta_3 t} \lambda_3^{j-1} \tag{2.79}$$

respectively, $j \geq 1$, $\forall t \geq 0$, where $||.||_p$ denotes any suitable vector norm. Then the series of (2.74)-(2.76) converge absolutely in the regions $\{|z| > \lambda_1,$ $\text{Re}\{s\} > \beta_1\}$, $\{|z| > \lambda_2, \text{Re}\{s\} > \beta_2\}$ and $\{|z| > \lambda_3, \text{Re}\{s\} > \beta_3\}$ respectively.

<u>Proof:</u> Consider $U(s,z)$ and note that

$$||U_j(s)||_p = ||\int_0^\infty e^{-st} U_j(t)dt||_p , \quad j \geq 1$$

$$\leq M_1 \int_0^\infty e^{(\beta_1 - s)t} \lambda_1^{j-1} dt$$

$$\leq \frac{M_1 \lambda_1^{j-1}}{\text{Re}\{(s-\beta_1)\}}, \qquad \text{Re}\{s\} > \beta_1 \tag{2.80}$$

i.e. $U_j(s)$ is well defined for $\text{Re}\{s\} > \beta_1$, $j \geq 1$. Now write

$U(s,z) = \sum_{j\geq 1} z^{1-j} U_j(s)$, then taking norms and using (2.77) yields

$$||U(s,z)||_p \leq \frac{M_1}{\text{Re}\{(s-\beta_1)\}} \sum_{j\geq 1} \left(\frac{\lambda_1}{|z|}\right)^{j-1}$$

$$= \frac{M_1}{\text{Re}\{(s-\beta_1)\}(1 - \frac{\lambda_1}{|z|})} \quad , \quad |z| > \lambda_1 \qquad (2.81)$$

and hence $U(s,z)$ is well defined, or converges absolutely, in the region $\{|z| > \lambda_1, \text{Re}\{s\} > \beta_1\}$. The proofs for $X(s,z)$ and $Y(s,z)$ follow in a similar manner and are hence omitted. ∎

Applying the 's transform' to (2.71)-(2.73) now yields

$$Y(s,z) = G(s,z)U(s,z) \qquad (2.82)$$

where $G(s,z)$ is the $m \times \ell$ 2D transfer-function matrix defined by

$$G(s,z) = (I_m - D(z))^{-1}C(sI_n - A - B(z)(I_m - D(z))^{-1}C)^{-1}\{B +$$

$$B(z)(I_m - D(z))^{-1}D_0\} + (I_m - D(z))^{-1}D_0 \qquad (2.83)$$

Hence application of the 'transforms' of definitions 2.5.1 and 2.5.2 respectively has resulted in an input/output description for (2.29) in the form of a 2D transfer-function matrix. Note also that (2.82)-(2.83) is invariant under the order of 'transform' application. In particular, it is easily shown that this description also results from first applying the 's-transform' of definition 2.5.2 to (2.29) and then applying the 'z-transform' of definition 2.5.1 to the result. Finally, this transfer-function matrix description extends in a natural manner to the case when (2.69) and/or (2.70) have non-zero entries and hence the details are omitted.

By appropriate rearrangement, it is possible to use $G(s,z)$ to obtain a clearer insight into the physical structure of (2.29). In particular, rewrite (2.82)-(2.83) in the form

$$Y(s,z) = G_0(s)U(s,z) + \sum_{j=1}^{M} G_j(s)z^{-j}Y(s,z) \qquad (2.84)$$

where

$$G_0(s) = C(sI_n - A)^{-1}B + D_0 \qquad (2.85)$$

and

$$G_j(s) = C(sI_n - A)^{-1}B_{j-1} + D_j, \quad 1 \leq j \leq M \qquad (2.86)$$

Then

$$G(s,z) = (I_m - \sum_{j=1}^{M} G_j(s)z^{-j})^{-1}G_0(s) \qquad (2.87)$$

and the block diagram structure of (2.84) is shown in Figure 2.8 or, equivalently, Figure 2.9. Either of these diagrams indicating that, in 2D transfer-function matrix terms, the dynamics of (2.29) are represented by a dynamic pre-compensator

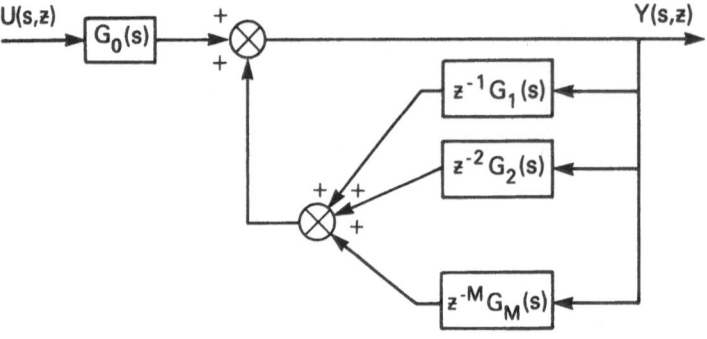

FIGURE 2.8

PRE-COMPENSATOR

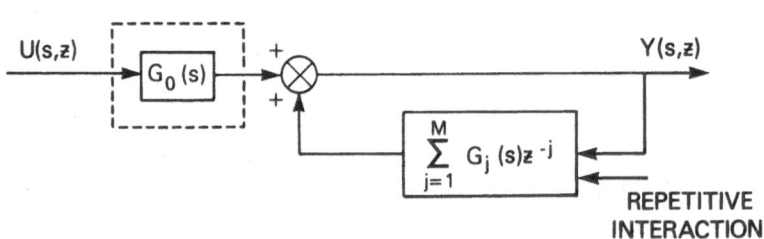

REPETITIVE
INTERACTION

FIGURE 2.9

followed by a positive feedback loop with unity gain in the forward path and dynamic elements in the feedback loop. These feedback elements are the repetitive interaction.

At this stage, suppose that $B_{j-1} = 0$, $D_j = 0$, $1 \leq j \leq M$, the pass subscript $k + 1$ is dropped, and the concept of a pass length is irrelevant. Then it follows immediately that $G(s,z)$ reduces to $G_0(s)$ of (2.85) which is just, see (2.51)-(2.52), the transfer-function matrix of the derived conventional linear system. Hence in this sense $G(s,z)$ can be regarded as the natural repetitive process generalisation of its well known and extensively used conventional linear systems counterpart.

In the unit memory case $G_1(s)$ of $G(s,z)$ is, see (2.54)-(2.55) with $M = 1$, the transfer-function matrix of the associated conventional linear system and hence, in effect, describes the contribution of Y_k to Y_{k+1}, $k \geq 0$. Consequently it is termed the interpass transfer-function matrix. For the more general non-unit memory case, element $G_j(s)$, $1 \leq j \leq M$, of (2.86) is, see (2.54)-(2.55), the transfer-function matrix of the jth associated conventional linear system and hence, in effect, describes the contribution of Y_{k+1-j} to Y_{k+1}, $k \geq 0$. Further, let $Y(s)$ denote the combined effects of the previous M passes. Then

$$Y(s) = \sum_{j=1}^{M} W^j(s) = \sum_{j=1}^{M} G_j(s)Y^{1-j}(s) \qquad (2.88)$$

and this expression can be interpreted in unit memory form by writing it as

$$\begin{bmatrix} Y^{2-M}(s) \\ \\ Y(s) \end{bmatrix} = \begin{bmatrix} 0 & I_m & & 0 \\ 0 & & 0 & I_m \\ G_M(s) & G_2(s) & G_1(s) \end{bmatrix} \begin{bmatrix} Y^{1-M}(s) \\ \\ Y^0(s) \end{bmatrix} \qquad (2.89)$$

Hence the non-unit memory version of the interpass transfer-function matrix is taken to be the following $N \times N$, $N = mM$, block companion matrix

$$G(s) = \begin{bmatrix} 0 & I_m & & 0 \\ 0 & & 0 & I_m \\ G_M(s) & G_2(s) & G_1(s) \end{bmatrix} \qquad (2.90)$$

The interpass transfer-function matrix will play a central role in subsequent chapters where use will also be made of the $N \times N$ constant coefficient block companion matrix defined as

$$D = \lim_{|s| \to +\infty} G(s) \qquad (2.91)$$

and hence

$$D = \begin{bmatrix} 0 & I_m & & 0 \\ 0 & & 0 & I_m \\ D_M & & D_2 & D_1 \end{bmatrix} \qquad (2.92)$$

For the discrete process of example 2.3.7, the 'z-transforms' are defined as follows.

Definition 2.5.3: The z-transforms of the sequences $U_{k+1}(P)$,

$X_{k+1}(P)$ and $Y_{k+1}(P)$, $0 \leq P \leq a$, $k \geq 0$ are defined by

$$U(P) = U_1(P) + z^{-1} U_2(P) + z^{-2} U_3(P) + \ldots \qquad (2.93)$$

$$X(P) = X_1(P) + z^{-1} X_2(P) + z^{-2} X_3(P) + \ldots \qquad (2.94)$$

and

$$Y(P) = Y_1(P) + z^{-1} Y_2(P) + z^{-2} Y_3(P) + \ldots \qquad (2.95)$$

respectively. ∎

A natural extension of the analysis of lemma 2.5.1 gives results on the convergence and existence properties of (2.93)-(2.95) and hence the details are omitted. Further, these series can be used in conjunction with the standard z-transform from conventional linear systems theory, termed the 'z_1-transform' in this context, to develop a 2D transfer-function matrix description of the state-space model (2.45). This invokes analysis which is just the natural extension of that used in the differential case and hence the details are omitted. The final result is

$$Y(z_1,z) = G(z_1,z) \ U(z_1,z) \qquad (2.96)$$

where

$$G(z_1,z) = (I_m - D(z))^{-1} C(z_1 I_n - \Phi - \Delta(z)(I_m - D(z))^{-1} C)^{-1} \{\Delta + $$
$$\Delta(z)(I_m - D(z))^{-1} D_0\} + (I_m - D(z))^{-1} D_0 \qquad (2.97)$$

with

$$\Delta(z) = \sum_{j=1}^{M} \Delta_{j-1} z^{-j}, \quad D(z) = \sum_{j=1}^{M} D_j z^{-j} \qquad (2.98)$$

or

$$G(z_1,z) = (I_m - \sum_{j=1}^{M} G_j(z_1) z^{-j})^{-1} G_0(z_1) \qquad (2.99)$$

where

$$G_0(z_1) = C(z_1 I_n - \Phi)^{-1} \Delta + D_0 \qquad (2.100)$$

and

$$G_j(z_1) = C(z_1 I_n - \Phi)^{-1} \Delta_{j-1} + D_j, \quad 1 \leq j \leq M \qquad (2.101)$$

This 2D transfer-function matrix description has an identical block diagram interpretation to that of Figures 2.8 or 2.9 for its differential counterpart. Further, the transfer-function matrices $G_0(z_1)$ and $G_j(z_1)$, $1 \leq j \leq M$, also have identical interpretations to their differential counterparts and the non-unit memory interpass transfer-function matrix is the following $N \times N$ block companion matrix

$$
G(z_1) = \begin{bmatrix} 0 & I_m & & 0 \\ & & & \\ 0 & 0 & & I_m \\ & & & \\ G_M(z_1) & G_2(z_1) & & G_1(z_1) \end{bmatrix} \tag{2.102}
$$

Use will also be made in the analysis of subsequent chapters of

$$
D = \lim_{|z_1| \to +\infty} G(z_1) \tag{2.103}
$$

which is just (2.92).

Notes and References

The industrial examples, unique control problem, and the abstract representation have evolved from the original work of Edwards (1974) and Owens (1977). Comprehensive details of the state-space model of (2.58) in section 2.4 can be found in Roesser (1975). For a comprehensive treatment of the transfer-function matrix descriptions see Rogers and Owens (1989a, 1990a). Edwards and Owens (1982) gives an extensive treatment of the classical stability analysis.

CHAPTER 3

STABILITY THEORY

A rigorous stability theory for the abstract representation of the linear constant pass length case given in definition 2.3.1 is developed. This is then applied to the special cases of examples 2.3.1, 2.3.3 and 2.3.7 to produce conditions suitable for the development of computationally feasible stability tests in chapter 4. In the same context, it is shown that an equivalence exists between stability of example 2.3.8 (the discrete unit memory case) and the well established area of stability analysis for 2D linear systems described by the Roesser model.

3.1 Asymptotic Stability

An illustrated by the simulation results of Figure 2.4, the essential unique undesirable feature of a repetitive process is the possible presence in the output sequence of oscillations which grow in amplitude from pass to pass. Hence the natural intuitive definition of asymptotic stability is to demand that, given any initial profile Y_0 and any disturbance sequence $\{b_k\}_{k \geq 1}$ which 'settles down' to a steady disturbance b_∞ as $k \to +\infty$, the sequence $\{Y_k\}_{k \geq 1}$ generated by $S(E_\alpha, W_\alpha, L_\alpha)$ 'settles down' to a steady profile Y_∞ as $k \to +\infty$. This idea is illustrated in Figure 3.1 and its major drawback is that it does not explicitly include the intuitive idea that asymptotic stability should be retained if the model is perturbed slightly due to modelling errors or simulation approximations. Consequently the following definition is preferred since it ensures that the 'set of stable systems' is open (in a well defined sense) in the class of all linear repetitive processes.

Definition 3.1.1: A linear repetitive process $S(E_\alpha, W_\alpha, L_\alpha)$ of constant pass length $\alpha > 0$ is said to be asymptotically stable if there exists a real scalar $\delta > 0$ such that, given any initial profile Y_0 and any strongly convergent disturbance sequence $\{b_k\}_{k \geq 1} \subset W_\alpha$, the sequence $\{Y_k\}_{k \geq 1}$ generated by the perturbed process

$$Y_{k+1} = (L_\alpha + \gamma)Y_k + b_{k+1}, \quad k \geq 0 \qquad (3.1)$$

converges strongly to a limit profile $Y_\infty \in E_\alpha$ whenever $||\gamma|| \leq \delta$.　■

Note: Y_∞ does, of course, depend on γ, Y_0 and $\{b_k\}_{k \geq 1}$.

The use of the term 'asymptotic stability' in the above definition can be justified by considering the case when $b_{k+1} = 0$, $k \geq 0$, which is strongly convergent to zero. Hence, from the definition, asymptotic stability requires that the

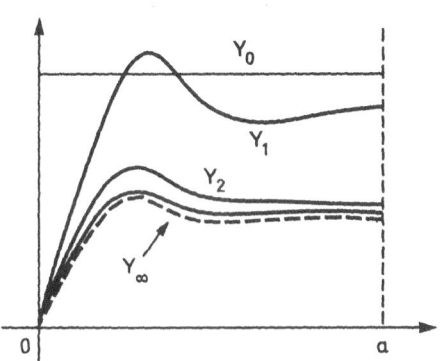

FIGURE 3.1

solution sequence $\{Y_k\}_{k\geq1}$ of (3.1) converges strongly to a profile Y_∞ for each Y_0 and model perturbation γ. In this particular case, the solution of (3.1) has the form

$$Y_k = (L_\alpha + \gamma)^k Y_0, \quad k \geq 0 \tag{3.2}$$

and set $\gamma = \dfrac{L_\alpha \delta}{||L_\alpha||}$ to yield $||\gamma|| = \delta$. Then it is immediately clear that the

sequence $(L_\alpha + \gamma)^k Y_0$, $k \geq 0$, is strongly convergent and hence bounded for each $Y_0 \in E_\alpha$. Application of the Banach-Steinhaus theorem now says that there exists a real number $M_\alpha > 0$ such that

$$||(L_\alpha + \gamma)^k|| \leq M_\alpha, \quad k \geq 0 \tag{3.3}$$

or, equivalently,

$$(1 + \frac{\delta}{||L_\alpha||})^k ||L_\alpha^k|| \leq M_\alpha, \quad k \geq 0 \tag{3.4}$$

Given (3.4), consider now the case of $\gamma = 0$ (the 'real' system) and define

$$\lambda_\alpha = (1 + \frac{\delta}{||L_\alpha||})^{-1} < 1 \tag{3.5}$$

In which case (3.4) takes the form

$$||L_\alpha^k|| \leq M_\alpha \lambda_\alpha^k, \quad k \geq 0 \tag{3.6}$$

and using (3.2) it follows that

$$||Y_k|| = ||L_\alpha^k Y_0|| \leq ||L_\alpha^k|| \; ||Y_0|| \leq M_\alpha \lambda_\alpha^k ||Y_0|| \tag{3.7}$$

Hence, in the absence of disturbances, the output sequence $\{Y_k\}_{k\geq1}$ converges strongly to zero for all initial profiles. Physically, this requires that the effects of the initial profile are rapidly attenuated after a 'large number of passes' at a geometric rate.

The key to constructing necessary and sufficient conditions for asymptotic stability of $S(E_\alpha, W_\alpha, L_\alpha)$ is to note its formal similarity with the matrix recursion relation which forms the basis of example 2.3.2. In particular, stability of the latter, viewed as a conventional linear system, is governed by the relationship between its eigenvalues/poles and the unit circle in the complex plane. Intuitively, therefore, it can be anticipated that asymptotic stability of $S(E_\alpha, W_\alpha, L_\alpha)$ will be related to the nature of the eigenvalues or, more generally, the spectral values of L_α which are defined as follows.

<u>Definition 3.1.2:</u> A complex number λ is said not to be a spectral value of L_α if, and only if, the bounded linear operator $\lambda I - L_\alpha$, where I is the identity operator

in E_α, has range dense in E_α and a bounded inverse $(\lambda I - L_\alpha)^{-1}$. The set, $\sigma(L_\alpha)$, of all spectral values of L_α is called the spectrum of L_α and its spectral radius is defined to be the finite positive number

$$r(L_\alpha) := \sup_{\lambda \in \sigma(L_\alpha)} |\lambda| \tag{3.8}$$

or, equivalently,

$$r(L_\alpha) = \lim_{k \to +\infty} ||L_\alpha^k||^{1/k} \tag{3.9}$$

∎

Theorem 3.1.1 below gives a necessary and sufficient condition for asymptotic stability of $S(E_\alpha, W_\alpha, L_\alpha)$ in terms of $r(L_\alpha)$ and its proof requires the following results. These are stated here without proof since they follow on straightforward use of standard results and concepts from functional analysis.

<u>Lemma 3.1.1</u>: Let ψ be a linear operator mapping E_α into itself with norm $||\psi|| < 1$. Then the operator $I - \psi$ has a bounded inverse which can be expressed in terms of the absolutely convergent power series.

$$(I - \psi)^{-1} = I + \psi + \psi^2 + \psi^3 + \ldots \tag{3.10}$$

and

$$||(I - \psi)^{-1}|| \leq \frac{1}{1 - ||\psi||} \tag{3.11}$$

∎

<u>Lemma 3.1.2</u>: If $r(L_\alpha) < 1$ and $r(L_\alpha) < \lambda < 1$, then

$$\beta(\lambda) := \sup_{|z| \geq \lambda} ||(zI - L_\alpha)^{-1}|| < +\infty \tag{3.12}$$

∎

<u>Lemma 3.1.3</u>: If $r(L_\alpha) < 1$ then there exists real scalars $\delta > 0$ and $r(L_\alpha) < \lambda < 1$ such that $r(L_\alpha + \gamma) < \lambda$ whenever $||\gamma|| \leq \delta$.

∎

<u>Lemma 3.1.4</u>: With the notation of lemma 3.1.3, suppose that $r(L_\alpha) < 1$ and $||\gamma|| \leq \delta$. Then there exists a real number $M_\alpha(\gamma) > 0$ such that

$$||(L_\alpha + \gamma)^k|| \leq M_\alpha(\gamma)\lambda^k, \quad k \geq 0 \tag{3.13}$$

∎

<u>Lemma 3.1.5</u>: With the above notation, suppose that $r(L_\alpha) < 1$ and $||\gamma|| \leq \delta$ and that the sequence $\{b_k\}_{k \geq 1}$ converges strongly to $b_\infty \in E_\alpha$. Then

$$\lim_{k \to +\infty} \sum_{j=1}^{k} (L_\alpha + \gamma)^{j-1}(b_{k+1-j} - b_\infty) = 0 \tag{3.14}$$

∎

(the limit being interpreted in the sense of the norm)

<u>Lemma 3.1.6</u>: With the above notation, suppose that $r(L_\alpha) < 1$ and $||\gamma|| \leq \delta$. Then the power series

$$\sum_{j=1}^{k} (L_\alpha + \gamma)^{j-1} b_\infty \tag{3.15}$$

is absolutely convergent as $k \to +\infty$.

∎

Given the above results, the following theorem can now be proved which gives a necessary and sufficient condition for asymptotic stability.

<u>Theorem 3.1.1:</u> The linear repetitive process $S(E_\alpha, W_\alpha, L_\alpha)$ of constant finite pass length $\alpha > 0$ is asymptotically stable if, and only if,

$$r(L_\alpha) < 1 \tag{3.16}$$

<u>Proof:</u> Suppose that $S(E_\alpha, W_\alpha, L_\alpha)$ is asymptotically stable. Then use of (3.5) and (3.6) in (3.9) yields

$$r(L_\alpha) = \lim_{k \to +\infty} ||L_\alpha^k||^{1/k} \leq \lambda_\alpha \lim_{k \to +\infty} M_\alpha^{1/k} = \lambda_\alpha < 1 \tag{3.17}$$

Conversely, suppose that $r(L_\alpha) < 1$ and write the solution of (3.1) in the form

$$Y_k = (L_\alpha + \gamma)^k Y_0 + \sum_{j=1}^{k} (L_\alpha + \gamma)^{j-1} b_{k+1-j}$$

$$= (L_\alpha + \gamma)^k Y_0 + \sum_{j=1}^{k} (L_\alpha + \gamma)^{j-1} (b_{k+1-j} - b_\infty)$$

$$+ \sum_{j=1}^{k} (L_\alpha + \gamma)^{j-1} b_\infty \tag{3.18}$$

Then, since $||(L_\alpha + \gamma)^k Y_0|| \leq M_\alpha(\gamma) \lambda^k ||Y_0||$ by lemma 3.1.4, it is clear that $(L_\alpha + \gamma)^k Y_0 \to 0$ as $k \to +\infty$. Further, it is easily verified using lemmas 3.1.5 and 3.1.6 that the sequence $\{Y_k\}_{k \geq 1}$ converges strongly to $Y_\infty \in E_\alpha$ where

$$\lim_{k \to +\infty} Y_k = Y_\infty := \sum_{j=1}^{\infty} (L_\alpha + \gamma)^{j-1} b_\infty \tag{3.19}$$

and the proof is complete. ∎

The result (3.16) provides a necessary and sufficient condition for asymptotic stability but little or no information concerning transient behaviour. The following definition and results are a physically realistic approach to characterising system behaviour after a 'large number of passes'.

<u>Definition 3.1.3:</u> Suppose that the linear repetitive process $S(E_\alpha, W_\alpha, L_\alpha)$ of constant pass length $\alpha > 0$ is asymptotically stable and let $\{b_k\}_{k \geq 1}$ be a disturbance sequence which converges strongly to a disturbance b_∞. Then the strong limit

$$Y_\infty := \lim_{k \to +\infty} Y_k \tag{3.20}$$

is termed the limit profile corresponding to $\{b_k\}_{k \geq 1}$. ∎

<u>Theorem 3.1.2:</u> Suppose that the linear repetitive process $S(E_\alpha, W_\alpha, L_\alpha)$ of constant pass length $\alpha > 0$ is asymptotically stable and let $\{b_k\}_{k \geq 1}$ be a disturbance sequence which converges strongly to a disturbance b_∞. Then the limit profile corresponding to this disturbance sequence is the unique solution of the linear equation.

$$Y_\infty = L_\alpha Y_\infty + b_\infty \qquad (3.21)$$

<u>Proof:</u> Use of (3.19) with $\gamma = 0$ yields a power series representation of Y_∞ which satisfies (3.21). The uniqueness of Y_∞ follows immediately on writing (3.21) in the form $(I - L_\alpha) Y_\infty = b_\infty$ and noting, by asymptotic stability, that $r(L_\alpha) < 1$ and hence $(I - L_\alpha)$ has a bounded inverse in E_α. Equivalently, (3.21) has a unique solution which can be written in the form

$$Y_\infty = (I - L_\alpha)^{-1} b_\infty \qquad (3.22)$$

<u>Corollary 3.1.2:</u> Y_∞ is independent of the initial pass profile Y_0 and of the direction of approach to b_∞. ∎

<u>Note:</u> Formally, (3.21) can be obtained from (3.1) with $\gamma = 0$ by replacing all variables by their strong limits.

    To be of use in a particular application, the abstract results of theorems 3.1.1 and 3.1.2 must be converted (if possible) into a suitable computational form. No general rules exist for this procedure, other than the obvious necessity to compute the spectral values of L_α and hence their moduli. Further, severe difficulties can arise if the space E_α and the operator L_α have a complex structure. For the special cases of examples 2.3.1, 2.3.3, 2.3.4, 2.3.7 and 2.3.8, however, the following analysis is possible. The approach to the spectral calculations used here is to consider the equation

$$(zI - L_\alpha)Y = \eta \qquad (3.23)$$

and construct necessary and sufficient conditions on the complex scalar z to ensure that a solution exists for all $\eta \in E_\alpha$, and that this solution is bounded in the sense that $||Y|| \leq K_0 ||\eta||$ for some real scalar $K_0 > 0$ and for all $\eta \in E_\alpha$. This yields the following results.

<u>Theorem 3.1.3:</u> A delay - algebraic system - The linear repetitive process of example 2.3.1 is asymptotically stable if, and only if,

$$|k_1| < 1 \qquad (3.24)$$

<u>Proof:</u> In this particular case (3.23) can be written in the form

$$zY(t) - W(t) = \eta(t)$$
$$W(t) = -k_0 W(t - X) + k_1 Y(t), \quad 0 \leq t \leq \alpha$$
$$W(t) = 0, \quad -X \leq t \leq 0 \qquad (3.25)$$

Further, by assumption, $X > 0$ and hence it is possible to choose an integer $n \geq 1$ such that $(n - 1) X < \alpha \leq n X$ and, after a little manipulation, (3.25) can be written in the matrix form

$$z \begin{bmatrix} Y(t) \\ Y(t + X) \\ Y(t + (n-1)X) \end{bmatrix} - \begin{bmatrix} W(t) \\ W(t + X) \\ W(t + (n-1)X) \end{bmatrix} = \begin{bmatrix} \eta(t) \\ \eta(t + X) \\ \eta(t + (n-1)X) \end{bmatrix} \quad (3.26)$$

Eliminating the W variables now gives

$$(zI_n - K) \begin{bmatrix} Y(t) \\ Y(t + X) \\ Y(t + (n-1)X) \end{bmatrix} = \begin{bmatrix} \eta(t) \\ \eta(t + X) \\ \eta(t + (n-1)X) \end{bmatrix}, \quad 0 \leq t < X \quad (3.27)$$

where the $n \times n$ matrix K is lower triangular with

$$K_{ii} = k_1, \; 1 \leq i \leq n, \; K_{i+1,j+1} = K_{i,j}, \; 1 \leq i,j \leq n - 1 \quad (3.28)$$

Choosing $z \neq k_1$, it is clear that (3.27) has a solution at each point $t \in [0,X]$ and it is easily shown that $Y(0) = 0$ and that $Y(t)$ is continuous on $0 \leq t \leq \alpha$. Further, consider, without loss of generality, the norm $||x||' = \max\limits_{1 \leq j \leq n} |x_j|$ in R^n and apply this to (3.27) to yield

$$||Y||' = ||(zI_n - K)^{-1}||' \cdot ||\eta||' \quad (3.29)$$

Hence the only candidate for a spectral value of L_α is $z = k_1$. In this case $(zI - K)$ is singular and it is a simple matter to construct η such that (3.27) has no solution. Hence $\sigma(L_\alpha) = \{k_1\}$, $r(L_\alpha) = |k_1|$, and the proof is complete. ∎

Suppose now that theorem 3.1.3 holds and a strongly convergent sequence $\{R_k\}_{k \geq 1}$ is applied. Then a representation of the corresponding limit profile $Y_\infty(t)$ can be obtained by, see theorem 3.1.2, replacing all variables in the defining equation (2.25) by their limiting values. This yields

$$Y_\infty(t) = -k_0 \, Y_\infty(t - X) + k_1 \, Y_\infty(t) + k_0 \, R_\infty(t), \; 0 \leq t \leq \alpha \quad (3.30)$$

or, after a little manipulation.

$$Y_\infty(t) = \frac{-k_0}{1 - k_1} Y_\infty(t - X) + \frac{k_0}{1 - k_1} R_\infty(t) \quad (3.31)$$

Theorem 3.1.4: The differential non-unit memory case - The differential non-unit memory linear repetitive process of example 2.3.3 is asymptotically stable if, and only if, all non-zero solutions of the relation

$$|zI_m - D_1 - z^{-1}D_2 - \ldots - z^{1-M}D_M| = 0 \quad (3.32)$$

have modulus strictly less than unity.

Proof: The abstract equation

$$(zI - L_\alpha) \begin{bmatrix} Y_1 \\ \vdots \\ Y_M \end{bmatrix} = \begin{bmatrix} \eta_1 \\ \vdots \\ \eta_M \end{bmatrix} \in E_\alpha^M \qquad (3.33)$$

in this particular case, where L_α has the block companion structure of (2.24) as defined by (2.32), can be written out as

$$z\,Y_j(t) = Y_{j+1}(t) + \eta_j(t), \quad 1 \le j \le M - 1 \qquad (3.34)$$

$$\dot{X}(t) = AX(t) + \sum_{j=1}^{M} B_{j-1} Y_{M+1-j}(t), \quad X(0) = 0$$

$$z\,Y_M(t) = CX(t) + \sum_{j=1}^{M} D_j Y_{M+1-j}(t) + \eta_M(t)$$

$$0 \le t \le \alpha \qquad (3.35)$$

Using induction, it is easily shown from (3.34) that

$$Y_p(t) = z^{p-M} Y_M(t) + \sum_{j=p}^{M-1} z^{p-j-1} \eta_j(t), \quad 1 \le p \le M - 1 \qquad (3.36)$$

and using this to remove $Y_1, Y_2, \ldots, Y_{M-1}$ from the differential equation in (3.35) yields

$$\dot{X}(t) = AX(t) + \sum_{j=1}^{M} B_{j-1} z^{1-j} Y_M(t) + \sum_{j=2}^{M} \sum_{i=M+1-j}^{M-1} B_{j-1} z^{M-j-i} \eta_i(t)$$

$$z\,Y_M(t) = CX(t) + \sum_{j=1}^{M} D_j z^{1-j} Y_M(t) + \sum_{j=2}^{M} \sum_{i=M+1-j}^{M-1} D_j z^{M-j-i} \eta_i(t) + \eta_M(t)$$

$$X(0) = 0, \quad 0 \le t \le \alpha \qquad (3.37)$$

Now define the polynomial matrix

$$P(z) = zI_m - D_1 - z^{-1}D_2 - \ldots - z^{1-M}D_M \qquad (3.38)$$

and suppose that $P(z)$ is nonsingular. Then (3.37) takes the form

$$\dot{X}(t) = (A + \sum_{j=1}^{M} B_{j-1} z^{1-j} P(z)^{-1} C)X(t)$$

$$+ \{\sum_{j=1}^{M} B_{j-1} z^{1-j}\} P(z)^{-1} \{\eta_M(t) + \sum_{i=1}^{M-1} \sum_{j=M+1-i}^{M} D_j z^{M-i-j} \eta_i(t)\}$$

$$+ \sum_{i=1}^{M-1} \sum_{j=M+1-i}^{M} B_{j-1} z^{M-i-j} \eta_i(t)$$

$$Y_M(t) = P(z)^{-1} \{CX(t) + \eta_M(t) + \sum_{i=1}^{M-1} \sum_{j=M+1-i}^{M} D_j z^{M-i-j} \eta_i(t)\} \qquad (3.39)$$

and it is clear in this case that $zI-L_\alpha$ has a bounded inverse and that the spectrum of L_α is hence a subset of the roots of $|P(z)| = 0$. Hence the proof will be complete if it can be shown that the spectrum of L_α is equal to the set of solutions of $|P(z)| = 0$. To achieve this set $t = 0$ in (3.37) and $\eta_1(t) = \ldots = \eta_{M-1}(t) \equiv 0$. In which case

$$P(z)Y_M(0) = \eta_M(0) \tag{3.40}$$

which has no solution if $\eta_M(0)$ is not in the range of $P(z)$, i.e. $zI - L_\alpha$ cannot be surjective if $|P(z)| = 0$. ∎

Corollary 3.1.4: The differential unit memory case-Setting $M = 1$ in theorem 3.1.4 gives the result that the differential unit memory linear repetitive process of example 2.3.4 is asymptotically stable if, and only if, all eigenvalues of the $m \times m$ matrix D_1 lie in the open unit circle in the complex plane. ∎

Note: It is rather surprising to find that asymptotic stability of the process of example 2.3.3 (and hence example 2.3.4) is independent of the system matrices and, in particular, independent of the eigenvalues of A. This is due entirely to the fact that α is finite and will change drastically when the case of $\alpha \to + \infty$ is considered in section 3.3.

Suppose now that theorem 3.1.4 holds and a strongly convergent sequence $\{U_k\}_{k \geq 1}$ is applied. Then a representation of the corresponding limit profile $Y_\infty(t)$ can, as for the case of theorem 3.1.3, be obtained by replacing all variables in the defining state-space model (2.29) by their strong limits. This yields

$$\dot{X}_\infty(t) = AX_\infty(t) + BU_\infty(t) + \hat{B}Y_\infty(t)$$
$$Y_\infty(t) = CX_\infty(t) + D_0U_\infty(t) + \hat{D}Y_\infty(t)$$
$$0 \leq t \leq \alpha, \quad X_\infty(0) = d_\infty \tag{3.41}$$

where

$$\hat{B} = \sum_{j=1}^{M} B_{j-1}, \quad \hat{D} = \sum_{j=1}^{M} D_j \tag{3.42}$$

or, since asymptotic stability ensures that $I_m - \hat{D}$ is nonsingular,

$$\dot{X}_\infty(t) = (A + \hat{B}(I_m - \hat{D})^{-1}C)X_\infty(t) + (B + \hat{B}(I_m - \hat{D})^{-1}D_0)U_\infty(t)$$
$$Y_\infty(t) = (I_m - \hat{D})^{-1}CX_\infty(t) + (I_m - \hat{D})^{-1}D_0U_\infty(t) \tag{3.43}$$

Setting $M = 1$ in (3.41) - (3.43) gives the corresponding results for the unit memory case.

Consider now asymptotic stability and the corresponding limit profiles for the discrete processes of examples 2.3.7 and 2.3.8. Then the following results are stated without proof since these follow from identical arguments to those used in establishing their differential counterparts.

Theorem 3.1.5: The discrete non-unit memory case - The discrete non-unit memory linear repetitive process of example 2.3.7 is asymptotically stable if, and only if, all non-zero solutions of (3.32) have modulus strictly less than unity. ∎

Corollary 3.1.5: The discrete unit memory case - The discrete unit memory linear repetitive process of example 2.3.8 is asymptotically stable if, and only if, all eigenvalues of the m×m matrix D_1 lie in the open unit circle in the complex plane. ∎

If theorem 3.1.5 holds, the corresponding limit profile state-space model is given by

$$X_\infty(P + 1) = (\Phi + \hat{\Delta}(I_m - \hat{D})^{-1}C)X_\infty(P) + (\Delta_0 + \hat{\Delta}(I_m - \hat{D})^{-1}D_0)U_\infty(P)$$

$$Y_\infty(P) = (I_m - \hat{D})^{-1}CX_\infty(P) + (I_m - \hat{D})^{-1}D_0U_\infty(P)$$

$$0 \leq P \leq \alpha , \quad X_\infty(0) = d_\infty \tag{3.44}$$

where

$$\hat{\Delta} = \sum_{j=1}^{M} \Delta_{j-1}, \quad \hat{D} = \sum_{j=1}^{M} D_j \tag{3.45}$$

Setting M = 1 in (3.44) - (3.45) gives the corresponding unit memory result.

Equations (3.43) and (3.44) are just the standard state-space models from differential and discrete conventional linear systems theory respectively. Hence if examples 2.3.3 and 2.3.7 (examples 2.3.4 and 2.3.8 in the unit memory case) are asymptotically stable then, in effect, their repetitive dynamics can, after a 'sufficiently large' number of passes, be described by a conventional linear systems state-space model. The implications of this result from a feedback control standpoint are discussed in chapter 6 which follows on from the stability related analysis of section 5.3.

3.2 Bounded-Input/Bounded-Output Stability

An alternative approach to stability analysis for $S(E_\alpha,W_\alpha,L_\alpha)$ is to demand that bounded disturbance sequences generate bounded sequences of pass profiles. Suppose also that this property is to be retained in the presence of small modelling errors. Then the following definition, compare with definition 3.1.1, is relevant.

Definition 3.2.1: A linear repetitive process $S(E_\alpha,W_\alpha,L_\alpha)$ of constant pass length $\alpha > 0$ is said to be bounded-input/bounded-output (BIBO) stable if there exists a real scalar $\delta > 0$ such that, given any Y_0 and $\{b_k\}_{k\geq1} \subset W_\alpha$ bounded in norm (i.e. $||b_k|| \leq c_1$ for some constant $c_1 \geq 0$ and for all k $\geq$ 1), the output sequence $\{Y_k\}_{k\geq1}$ generated by the perturbed process (3.1) is bounded in norm whenever $||\gamma|| \leq \delta$. ∎

The following theorem establishes the equivalence of the, apparently different, definitions of 3.1.1 and 3.2.1.

<u>Theorem 3.2.1</u>: The linear repetitive process $S(E_\alpha, W_\alpha, L_\alpha)$ of constant pass length $\alpha > 0$ is BIBO stable if, and only if, it is asymptotically stable.

<u>Proof</u>: Suppose that $S(E_\alpha, W_\alpha, L_\alpha)$ is BIBO stable. Then consideration of the bounded sequence $b_k = 0$, $k \geq 1$, and $\gamma = \dfrac{\delta L_\alpha}{||L_\alpha||}$ leads to the boundedness of (3.2) and hence to (3.6) where λ_α is defined by (3.5). It now follows from (3.17) and theorem 3.1.1 that the system is asymptotically stable.

Conversely, suppose that $S(E_\alpha, W_\alpha, L_\alpha)$ is asymptotically stable (and hence $r(L_\alpha) < 1$) and that the disturbance sequence is bounded, i.e. $||b_k|| \leq c_1$, $k \geq 1$. Further, write the solution of (3.1) in the form

$$Y_k = (L_\alpha + \gamma)^k Y_0 + \sum_{j=1}^{k} (L_\alpha + \gamma)^{j-1} b_{k+1-j} \tag{3.46}$$

Then taking the norm of (3.46) and making use of the result of lemma 3.1.4 yields

$$||Y_k|| \leq ||(L_\alpha + \gamma)^k|| \, ||Y_0|| + \sum_{j=1}^{k} ||(L_\alpha + \gamma)^{j-1}|| \, ||b_{k+1-j}||$$

$$\leq M_\alpha(\gamma) \lambda^k (||Y_0|| + \frac{c_1}{1-\lambda})$$

$$\leq M_\alpha(\gamma) (||Y_0|| + \frac{c_1}{1-\lambda}) \tag{3.47}$$

whenever $||\gamma|| \leq \delta$. It now follows immediately that $S(E_\alpha, W_\alpha, L_\alpha)$ is BIBO stable. ∎

3.3 Stability Along the Pass

Under well defined conditions, asymptotic stability of $S(E_\alpha, W_\alpha, L_\alpha)$ in the form of theorem 3.1.1 guarantees the existence of a limit profile. It does not, however, also guarantee that this limit profile has acceptable dynamic characteristics. To illustrate this point, consider the following SISO differential unit memory process where β is a real scalar

$$\dot{X}_{k+1}(t) = -X_{k+1}(t) + U_{k+1}(t) + (1 + \beta)Y_k(t)$$
$$Y_{k+1}(t) = X_{k+1}(t)$$
$$0 \leq t \leq \alpha, \quad X_{k+1}(0) = 0, \quad k \geq 0 \tag{3.48}$$

Then, since $D_1 = 0$, use of corollary 3.1.4 immediately yields asymptotic stability and substitution in (3.43) with $M = 1$ then gives the corresponding limit profile description as

$$\dot{Y}_\infty(t) = \beta \, Y_\infty(t) + U_\infty(t)$$
$$0 \le t \le \alpha, \quad Y_\infty(0) = 0 \tag{3.49}$$

Further, it is easily shown that if $U_{k+1}(t) \equiv 1$, $Y_0(t) \equiv 0$, $0 \le t \le \alpha$, $k \ge 0$, then

$$Y_1(t) = 1 - e^{-t}, \quad 0 \le t \le \alpha \tag{3.50}$$

and

$$Y_\infty(t) = \frac{1}{\beta}(e^{\beta t} - 1), \quad 0 \le t \le \alpha \tag{3.51}$$

Hence, despite the fact that the first pass profile $Y_1(t)$ is a quite acceptable 'classical' response to the unit step demand $U_1(t) \equiv 1$, the limit profile can have totally unacceptable dynamic characteristics. For example, if $\beta > 0$ then the limit profile grows exponentially and can be said to be 'unstable along the pass' in an obvious intuitive sense.

The natural definition of stability along the pass for cases such as that highlighted by the above analysis is to demand that the limit profile is stable in the standard, or conventional, sense as $\alpha \to +\infty$ (i.e. $\beta < 0$ in (3.48)). Unfortunately, however, this intuitively appealing idea does not apply in a simple manner to processes such as those described by example 2.3.5. Consequently the following analysis develops the concept of stability along the pass by considering the rate of approach to the limit profile as $\alpha \to +\infty$.

Suppose that that $S(E_\alpha, W_\alpha, L_\alpha)$ can be modelled over the range of pass lengths $\alpha \ge \alpha_0$, where α_0 is some nominal value of interest, and introduce the following terminology.

<u>Definition 3.3.1:</u> A collection of models of $S(E_\alpha, W_\alpha, L_\alpha)$ with pass lengths in the range $\alpha \ge \alpha_0$ is termed an extended linear repetitive process and is denoted $S(E_\alpha, W_\alpha, L_\alpha)_{\alpha \ge \alpha_0}$. ∎

Further, suppose that a particular element of $S(E_\alpha, W_\alpha, L_\alpha)_{\alpha \ge \alpha_0}$ is asymptotically stable. Then the following result shows that this element can be partially characterised by real scalars $M_\alpha > 0$ and $0 < \lambda_\alpha < 1$ describing the rate of approach to the limit profile.

<u>Theorem 3.3.1:</u> Suppose that the linear repetitive process $S(E_\alpha, W_\alpha, L_\alpha)$ of constant pass length $\alpha > 0$ is asymptotically stable. Further, let this process be subjected to a constant disturbance sequence $b_{k+1} = b_\infty$, $k \ge 0$, which generates the limit profile Y_∞. Then there exists real scalars $M_\alpha > 0$ and $0 < \lambda_\alpha < 1$ such that

$$||Y_k - Y_\infty|| \leq M_\alpha \lambda_\alpha^k \{||Y_0|| + \frac{||b_\infty||}{1 - \lambda_\alpha}\}, \quad k \geq 0 \qquad (3.52)$$

Proof: Since $b_{k+1} = b_\infty$, $k \geq 0$, the solution of (3.1) with $\gamma = 0$ can be written as

$$Y_k = L_\alpha^k Y_0 + \sum_{j=1}^{k} L_\alpha^{j-1} b_\infty \qquad (3.53)$$

and, using (3.19) with $\gamma = 0$, the limit profile can be expressed as

$$Y_\infty = \sum_{j=1}^{\infty} L_\alpha^{j-1} b_\infty \qquad (3.54)$$

Hence the 'error' $Y_k - Y_\infty$ can be written in the form

$$Y_k - Y_\infty = L_\alpha^k Y_0 - \sum_{j=k+1}^{\infty} L_\alpha^{j-1} b_\infty, \quad k \geq 0 \qquad (3.55)$$

or, taking the norm to obtain a (numerical) estimate of convergence,

$$||Y_k - Y_\infty|| \leq ||L_\alpha^k|| \; ||Y_0|| + \sum_{j=k+1}^{\infty} ||L_\alpha^{j-1}|| \; ||b_\infty|| \qquad (3.56)$$

The proof is now completed by using (3.6). ∎

Note: In effect, this result states that the output sequence $\{Y_k\}_{k \geq 1}$ approaches the limit profile at a geometric rate governed by λ_α. This result, together with definition 3.3.2 below, plays a significant part in the analysis of chapter 5.

Given theorem 3.3.1, the following definition of stability along the pass is expressed in terms of the existence of finite bounds on the scalars M_α and λ_α as $\alpha \to +\infty$. Its effective action is to demand that the rate of approach of the output sequence to the limit profile has a guaranteed geometric upper bound independent of pass length for $\alpha \geq \alpha_0$.

Definition 3.3.2: The extended linear repetitive process $S(E_\alpha, W_\alpha, L_\alpha)_{\alpha \geq \alpha_0}$ is said to be stable along the pass if there exists finite real scalars $M_\infty > 0$ and $0 < \lambda_\infty < 1$ such that, for each $\alpha \geq \alpha_0$ and for each constant disturbance sequence $b_{k+1} = b_\infty$, $k \geq 0$, the output sequence from the model $S(E_\alpha, W_\alpha, L_\alpha)$ satisfies the inequality

$$||Y_k - Y_\infty|| \leq M_\infty \lambda_\infty^k \{||Y_0|| + \frac{||b_\infty||}{1 - \lambda_\infty}\}, \quad k \geq 0 \qquad (3.57)$$

∎

Despite its well defined physical meaning, this definition is not in appropriate form for the derivation of stability criteria. A more useful definition is implied by lemma 3.3.1 below which leads to the central result of this section in the form of theorem 3.3.2.

Lemma 3.3.1: The extended linear repetitive process $S(E_\alpha, W_\alpha, L_\alpha)_{\alpha \geq \alpha_0}$ is stable along

the pass if, and only if, there exists finite real numbers $M_\infty > 0$ and

$0 < \lambda_\infty < 1$ such that

$$||L_\alpha^k|| \leq M_\infty \lambda_\infty^k \qquad (3.58)$$

for all $\alpha \geq \alpha_0$.

Proof: Suppose that $S(E_\alpha, W_\alpha, L_\alpha)_{\alpha \geq \alpha_0}$ is stable along the pass and set $b_\infty = 0$ in

(3.57) where $Y_\infty = 0$ is the corresponding limit profile. Then this inequality

reduces to $||Y_k|| \leq M_\infty \lambda_\infty^k ||Y_0||$ for all $\alpha \geq \alpha_0$ and, since Y_0 is arbitrary and

$Y_k = L_\alpha^k Y_0$, (3.58) follows trivially.

Conversely, suppose that (3.58) holds. Then (3.57) follows in a similar manner
to the proof of theorem 3.3.1, i.e. the extended process is stable along the pass. ∎

Theorem 3.3.2: The extended linear repetitive process $S(E_\alpha, W_\alpha, L_\alpha)_{\alpha \geq \alpha_0}$ is stable

along the pass if, and only if,

(a) $r_\infty := \sup_{\alpha \geq \alpha_0} r(L_\alpha) < 1$ $\qquad (3.59)$

and

(b) $M_0 := \sup_{\alpha \geq \alpha_0} \sup_{|z| \geq \lambda} ||(zI - L_\alpha)^{-1}|| < +\infty$ $\qquad (3.60)$

for some real number $\lambda \in (r_\infty, 1)$.

Proof: To prove necessity, note from lemma 3.3.1 that

$$r(L_\alpha) = \lim_{k \to +\infty} ||L_\alpha^k||^{1/k} \leq \lim_{k \to +\infty} \lambda_\infty M_\infty^{1/k} = \lambda_\infty < 1 \qquad (3.61)$$

for all $\alpha \geq \alpha_0$ and hence $r_\infty \leq \lambda_\infty < 1$. Suppose also that λ is any

number in the range $\lambda_\infty < \lambda < 1$. Then it is clear that $(zI - L_\alpha)^{-1}$ can be expressed

as an absolutely convergent power series in z^{-1} for $|z| \geq \lambda$ and

$$||(zI - L_\alpha)^{-1}|| = |z^{-1}| \; ||\sum_{j=1}^{\infty} \frac{1}{z^j} L_\alpha^j||$$

$$\leq \frac{1}{\lambda} \sum_{j=0}^{\infty} \frac{1}{\lambda^j} ||L_\alpha^j||$$

$$\leq \frac{1}{\lambda} \sum_{j=0}^{\infty} (\frac{\lambda_\infty}{\lambda})^j M_\infty$$

$$= \frac{M_\infty}{\lambda(1 - \lambda/\lambda_\infty)} \tag{3.62}$$

for all $\alpha \geq \alpha_0$.

For a proof of sufficiency, suppose that $r_\infty < 1$ and consider the contour M in the complex plane defined by the relation

$$M = \{z: |z| = \lambda\} \tag{3.63}$$

Further, exploit standard results from operational calculus in Banach spaces to write L_α^k as the contour integral

$$L_\alpha^k = \frac{1}{2\pi i} \int_M z^k (zI - L_\alpha)^{-1} dz \tag{3.64}$$

where $z = \lambda e^{i\theta}$ and i denotes the 'square root of minus one'. Then taking norms and using (3.60) yields.

$$||L_\alpha^k|| \leq \frac{1}{2\pi} || \int_0^{2\pi} \lambda^k e^{ik\theta} (\lambda e^{i\theta} - L_\alpha)^{-1} \lambda i e^{i\theta} d\theta ||$$

$$\leq \frac{1}{2\pi} \int_0^{2\pi} \lambda^{k+1} ||(\lambda e^{i\theta} - L_\alpha)^{-1}|| d\theta$$

$$\leq (M_0 \lambda) \lambda^k \tag{3.65}$$

This verifies (3.58) with $M_\infty = M_0 \lambda$ and $\lambda = \lambda_\infty$ and the proof is complete. ∎

Note 1: It can be shown that (3.60) can be relaxed slightly to

$$M_0 := \sup_{\alpha \geq \alpha_0} \ \sup_{|z| = \lambda} ||(zI - L_\alpha)^{-1}|| < +\infty \tag{3.66}$$

Note 2: (3.59) is equivalent to asymptotic stability for all $\alpha \geq \alpha_0$. Hence the reason for retaining the separate identities of (a) and (b) in theorem 3.3.2, despite the fact that (b) does imply (a).

The 'boundedness' condition (b) of theorem 3.3.2 is equivalent to the existence of a $\lambda \in (r_\infty, 1)$ such that (3.23), i.e.

$$(zI - L_\alpha)Y = \eta \tag{3.67}$$

has a uniformly bounded, with respect to α, solution $Y \in E_\alpha$ for all choices of $\eta \in E_\alpha$ satisfying $\sup_\alpha ||\eta|| < +\infty$ and for all $|z| \geq \lambda$. Further, it is clear that, in general, this condition could prove very difficult to interpret. For the special cases of examples 2.3.1, 2.3.3,, 2.3.4, 2.3.7 and 2.3.8, however, the following results are obtained.

Theorem 3.3.3: A delay-algebraic system - The extended linear repetitive process $S(E_\alpha, W_\alpha, L_\alpha)_{\alpha \geq \alpha_0}$ generated by (2.25) of example 2.3.1 with $\alpha \geq \alpha_0$ is stable along the pass if, and only if,

(a) $|k_1| < 1$ (3.68)

and

(b) $|k_0| < 1 - |k_1|$ (3.69)

Proof: In this particular case the extended process simply consists of the family of models (2.25) with $\alpha \geq \alpha_0$ for some α_0 of interest. Further, it follows immediately from theorem 3.1.3 that $r(L_\alpha) = |k_1|$ for all $\alpha > 0$ and hence (a) of theorem 3.3.2 in this particular case reduces to (3.68).

Considering now (b) of theorem 3.3.2, then, in this particular case, (3.67) can be written as

$$Y(t) = \frac{-z\, k_0}{z - k_1} Y(t - X) + \frac{(k_0 \eta(t - X) + \eta(t))}{z - k_1} \qquad (3.70)$$

Hence it follows immediately that (3.60) holds if, and only if,

$$\sup_{|z| \geq \lambda} \left| \frac{-z\, k_0}{z - k_1} \right| < 1 \qquad (3.71)$$

for some $\lambda \in (|k_1|, 1)$. Elementary graphical considerations now reduce (3.71) to (3.69) and the proof is complete. ∎

In the above theorem, (3.68) is a consequence of (3.69). Hence testing example 2.3.1 for stability along the pass reduces to the very simple task of checking (3.69). Consequently no further consideration of this example is undertaken here except to note the analysis below which generalises theorem 3.3.3 to non-unit memory processes with a current pass delay such as bench mining systems. Here the dynamics are described by

$$Y_{k+1}(t) = -k_0 Y_{k+1}(t - X) + \sum_{j=1}^{M} k_j Y_{k+1-j}(t) + k_0 R_{k+1}(t) \qquad 0 \leq t \leq \alpha, \quad k \geq 0$$

$$Y_{k+1}(t) = 0, \quad -X \leq t \leq 0 \qquad (3.72)$$

where k_0 and k_j, $1 \leq j \leq M$, are real constants of arbitrary sign. Further, (3.72) reduces to the equation of example 2.3.1 if $M = 1$ and hence in this sense it can be regarded as the natural generalisation of this unit memory process. Suppose also that $E_\alpha = W_\alpha$ is chosen as the vector space of continuous functions on $[0, \alpha]$ with initial condition $Y(0) = 0$ and norm $||Y|| = \max_{0 \leq t \leq \alpha} |Y(t)|$. Then (3.72) can be written in the abstract form $S(E_\alpha, W_\alpha, L_\alpha)$, where L_α has the 'companion form' of (2.24) and L_α^j, $1 \leq j \leq M$, is defined by considering

$$Y_1(t) = -k_0 Y_1(t - X) + k_j Y_{1-j}(t), \ 0 \leq t \leq \alpha$$

$$Y_1(t) = 0, \ -X \leq t \leq 0 \tag{3.73}$$

The following result now provides necessary and sufficient conditions for stability along the pass of (3.72) and generalises theorem 3.3.3 to this case.

Theorem 3.3.4: The extended linear repetitive process $S(E_\alpha, W_\alpha, L_\alpha)_{\alpha \geq \alpha_0}$ generated by (3.72) with $\alpha \geq \alpha_0$ is stable along the pass if, and only if,

(a) $\sup\{|z|: \ z^M - \sum\limits_{j=1}^{M} k_j \ z^{M-j} = 0\} < 1 \tag{3.74}$

and

(b) $\max\limits_{|z|=1} |f(z)| < 1 \tag{3.75}$

where

$$f(z) = \cfrac{- k_0}{1 - \sum\limits_{j=1}^{M} k_j z^{-j}} \tag{3.76}$$

Proof: The first step in proving (a) is to compute the spectral values of L_α which, since $\alpha > 0$ is finite, are simply the eigenvalues. Hence the computation of $r(L_\alpha)$ reduces to finding the complex scalar of largest modulus which satisfies the eigenvalue problem

$$L_\alpha W = zW \in E_\alpha^M \tag{3.77}$$

Writing out (3.77) yields

$$zW_i(t) = W_{i+1}(t), \ 1 \leq i \leq M - 1 \tag{3.78}$$

$$zW_M(t) = -k_0 z \ W_M(t - X) + \sum\limits_{j=1}^{M} k_j W_{M+1-j}(t) \tag{3.79}$$

$$0 \leq t \leq \alpha$$

or

$$W_1(t) = \cfrac{-z^M k_0 W_1(t - X)}{z^M - \sum\limits_{j=1}^{M} k_j z^{M-j}} \tag{3.80}$$

on using (3.78) to write

$$W_i(t) = z^{i-1} W_1(t), \ 2 \leq i \leq M \tag{3.81}$$

and appropriate substitution in (3.79). Hence, noting the initial conditions associated with (3.73), it follows immediately that the non-trivial solutions of (3.77) are the roots of

$$z^M - \sum\limits_{j=1}^{M} k_j \ z^{M-j} = 0 \tag{3.82}$$

and consequently

$$r(L_\alpha) = \sup\{|z|: \quad z^M - \sum_{j=1}^{M} k_j z^{M-j} = 0\} \tag{3.83}$$

The result now follows immediately on noting that $r(L_\alpha)$ is independent of α.

To prove (b), first note the interpretation of (b) of the general result of theorem 3.3.2 in terms of (3.67) and write out the abstract equation

$$(zI - L_\alpha) \begin{bmatrix} Y_1 \\ \\ Y_M \end{bmatrix} = \begin{bmatrix} \eta_1 \\ \\ \eta_M \end{bmatrix} \in E_\alpha^M \tag{3.84}$$

for this case over $0 \le t \le \alpha$ as

$$z\, Y_i(t) - Y_{i+1}(t) = \eta_i(t) , \quad 1 \le i \le M - 1 \tag{3.85}$$

$$z\, Y_M(t) - W(t) = \eta_M(t) \tag{3.86}$$

$$W(t) = -k_0 W(t - X) + \sum_{j=1}^{M} k_j Y_{M+1-j}(t) \tag{3.87}$$

$$W(t) = 0 , \quad -X \le t \le 0 \tag{3.88}$$

or

$$z\, Y_i(t) - Y_{i+1}(t) = \eta_i(t) , \quad 1 \le i \le M - 1 \tag{3.89}$$

$$z\, Y_M(t) = -k_0 z\, Y_M(t - X) + \sum_{j=1}^{M} k_j\, Y_{M+1-j}(t)$$
$$+ k_0\, \eta_M(t - X) + \eta_M(t) \tag{3.90}$$

Using induction, it now follows from (3.89) that

$$Y_i(t) = z^{i-1}\, Y_1(t) - \sum_{j=1}^{i-1} z^{i-j-1} \eta_j(t), \quad 2 \le i \le M \tag{3.91}$$

and use of this in (3.90) yields

$$Y_1(t) = f(z) Y_1(t - X) + \frac{1}{\rho(z)} [k_0 \sum_{p=1}^{M-1} z^{-p} \eta_p(t - X)$$
$$- \sum_{j=1}^{M} \sum_{p=1}^{M-j} z^{-p-j} \eta_p(t) + k_0 \eta_M(t - X) + \eta_M(t)] \tag{3.92}$$

where

$$\rho(z) = 1 - \sum_{j=1}^{M} k_j\, z^{-j} \tag{3.93}$$

A simple argument now indicates that the existence of a uniform bound for $Y_1(t)$ over the required ranges of α and z is equivalent to the existence of a uniform bound for the solutions of (3.84) over these same ranges for all permissible choices of η_i, $1 \le i \le M$. Consequently it remains to prove that the existence of such a bound is equivalent to the validity of (3.75) with $|z| = 1$ replaced by $|z| \ge \lambda$.

The fact that this condition is sufficient follows immediately on noting that the terms in (3.92) arising from $\eta_p(t - X)$ and $\eta_p(t)$ are uniformly bounded $\forall |z| \ge \lambda$ $\in (r_\infty, 1)$. To show necessity, first note that $f(z)$ is analytic $\forall |z| \ge \lambda$ and all λ

$\epsilon(r_\infty,1)$. Further, (a) is equivalent to $\rho(z) \neq 0$, $\forall\ |z| \geq 1$, and it follows immediately that the supremum of $|f(z)|$ occurs at a finite value, say, $z = z^*$. Now set $z = z^*$ in (3.92) and note that it is always possible to choose η_i, $1 \leq i \leq M$, such that all terms in the resulting version of this equation which do not involve Y_1 are nonzero and constant. In which case a contradiction occurs if $|f(z^*)| \geq 1$ since (3.92) is then unstable and hence $Y_1(t)$ is not uniformly bounded on $[0,+\infty]$.

At this stage, it has been shown that stability along the pass is equivalent to the existence of a $\lambda \in (r_\infty,1)$ such that

$$\sup_{|z|\geq\lambda}\ |f(z)| < 1 \qquad\qquad (3.94)$$

Further, the proof to date has shown that the supremum of $f(z)$ occurs at a finite value of z and hence, by the maximum modulus theorem, at the boundary, i.e. $|z| \geq \lambda$ can be replaced by the compact set $|z| = \lambda$. The proof is now completed by showing that (3.94) is a necessary and sufficient condition for (3.75) where the latter is obvious. To establish the former, suppose that

$$\sup_{|z|=1}\ |f(z)| < 1 \qquad\qquad (3.95)$$

Then continuity implies that for each $|z| = 1$, there exists $r(z)$ such that

$$|f(z^1)| < 1 \qquad\qquad (3.96)$$

for $|z^1 - z| < r(z)$. A simple argument based on the compactness of $|z| = 1$ now yields the existence of a λ such that (3.94) holds. ∎

The following corollaries yield useful information concerning the testing of (a) and/or (b) of theorem 3.3.4 for a given example.

Corollary 3.3.4: A necessary condition for (a) of theorem 3.3.4 is that $\rho(z)$ of (3.93) satisfies $\rho(1) > 0$ and hence

$$1 - \sum_{j=1}^{M} k_j > 0 \qquad\qquad (3.97)$$

Proof: First note again that (3.74) is equivalent to

$$\rho(z) \neq 0, \quad \forall\ |z| \geq 1 \qquad\qquad (3.98)$$

Now consider positive real values of z and suppose that

$$1 - \sum_{j=1}^{M} k_j \leq 0 \qquad\qquad (3.99)$$

Then this contradicts (3.98) since it is easily seen that there exists a positive real number $z^* \geq 1$ such that $\rho(z^*) = 0$. ∎

Corollary 3.3.5: Suppose (as in bench mining systems) that $k_j > 0$, $1 \leq j \leq M$. Then $\rho(1) > 0$ is a necessary and sufficient condition for (a) of theorem 3.3.4.
Proof: If (a) holds then $\rho(1) > 0$ is immediate from corollary 3.3.4. For the converse, it is required to prove that if (3.97) holds then $\rho(z) \neq 0$, $\forall\ |z| \geq 1$.

This follows immediately on noting that

$$\left| \sum_{j=1}^{M} k_j z^{-j} \right| \leq \sum_{j=1}^{M} k_j < 1, \quad \forall \ |z| \geq 1 \tag{3.100}$$

and hence

$$|\rho(z)| = \left| 1 - \sum_{j=1}^{M} k_j z^{-j} \right| \geq 1 - \sum_{j=1}^{M} k_j > 0, \quad \forall \ |z| \geq 1 \tag{3.101}$$

■

Corollary 3.3.6: Suppose (as in bench mining systems) that $k_0 > 0$ and $k_j > 0$, $1 \leq j \leq M$. Then

$$k_0 < 1 - \sum_{j=1}^{M} k_j \tag{3.102}$$

is a necessary and sufficient condition for (a) and (b) of theorem 3.3.4.

Proof: Here it is required to prove that (3.102) is equivalent to both of these conditions. To prove the first, simply note that for (3.102) to hold $1 - \sum_{j=1}^{M} k_j > 0$, which is simply the result of corollary 3.3.4. For the second, a simple argument yields

$$\max_{|z|=1} |f(z)| = \frac{k_0}{1 - \sum_{j=1}^{M} k_j} \tag{3.103}$$

■

Corollary 3.3.7: A sufficient condition for (b) of theorem 3.3.4 is that

$$|k_0| < 1 - \sum_{j=1}^{M} |k_j| \tag{3.104}$$

Proof: First note from (3.104) that $\sum_{j=1}^{M} |k_j| < 1$ and hence

$$\left| \sum_{j=1}^{M} k_j z^{-j} \right| \leq \sum_{j=1}^{M} |k_j| < 1, \quad \forall \ |z| = 1 \tag{3.105}$$

Using (3.105) now yields

$$\left| 1 - \sum_{j=1}^{M} k_j z^{-j} \right| \geq 1 - \sum_{j=1}^{M} |k_j|, \ \forall \ |z| = 1 \tag{3.106}$$

and the result follows immediately since

$$\max_{|z|=1} |f(z)| \leq \frac{|k_0|}{1 - \sum_{j=1}^{M} |k_j|} < 1 \tag{3.107}$$

■

Testing (a) of theorem 3.3.4 in the general case is easily converted into the standard stability problem from discrete conventional linear systems theory. In the case of (b), let ❦ denote the closed contour generated by $\rho(z)$ as z traverses the unit circle in the complex plane either clockwise or anti-clockwise. Then this condition holds if, and only if, ❦ lies completely outside the circle in the complex plane of radius $|k_0|$ and centre the origin. This again is easily tested using elements of conventional linear systems stability analysis.

Considering now the processes of examples 2.3.3, 2.3.4, 2.3.7 and 2.3.8 yields the following results. Further, those for the discrete processes of examples 2.3.7 and 2.3.8 are stated without proofs since these follow from identical arguments to those used in establishing their differential counterparts.

Theorem 3.3.5: The differential non-unit memory case - Suppose that

   (i)   the pair $\{C,A\}$ is observable,

   (ii) the pair $\{A, \sum_{j=1}^{M} B_{j-1}\gamma^{j-1}\}$ is controllable at all but a finite

         number of points $\gamma_1, \gamma_2, \ldots, \gamma_q$ in the complex plane,

and

   (iii) $|sI_n - A - \sum_{j=1}^{M} B_{j-1}\gamma_i^{j-1}P(\gamma_i^{-1})^{-1}C|$ has no roots on the imaginary axis

         of the complex plane, $1 \leq i \leq q$, where $P(\gamma)$ is defined by (3.38) with z
         replaced by γ.

Then the extended linear repetitive process $S(E_\alpha, W_\alpha, L_\alpha)_{\alpha \geq \alpha_0}$ generated by

differential models of the form (2.29) in example 2.3.3 with $\alpha \geq \alpha_0$ is stable along

the pass if, and only if,

   (a) $r_\infty = \sup\{|z| : P(z) = 0\} < 1$ (3.108)

and

   (b)   there exists real numbers $\epsilon > 0$ and $r_\infty < \lambda < 1$ such that

         $$|sI_n - A - \sum_{j=1}^{M} B_{j-1}z^{1-j}P(z)^{-1}C| \neq 0$$ (3.109)

         for all complex numbers z, s satisfying $|z| \geq \lambda$ and $\mathrm{Re}\{s\} \geq -\epsilon$.

Proof: To prove (a), note from theorem 3.1.4 that $r(L_\alpha)$ is independent of α and

hence the result. In the case of (b), first note again that (b) of the general
result, theorem 3.3.2, is equivalent to the existence of the uniform bound defined
in terms of (3.67) which, in this particular case, is equivalent to the existence of
such a bound on the solutions of (3.34) - (3.35) as $\alpha \to +\infty$. Given (3.36), this is
equivalent to the existence of such a bound on the solutions of (3.39) as $\alpha \to +\infty$.
Further, this last requirement is equivalent to (b) above since the fact that all of
the coefficients of $\eta_1, \eta_2, \ldots, \eta_M$ in (3.39) are bounded in any region $|z| \geq \lambda$ with

$\lambda > r_\infty$ reduces the boundedness condition to a stability condition on the matrix

$$\Psi(z) = A + \sum_{j=1}^{M} B_{j-1}z^{1-j}P(z)^{-1}C$$ (3.110)

for $|z| \geq \lambda$. The fact that this must be a strong stability condition follows from
the following argument based on the controllability and observability assumptions.

Suppose that it is only possible to choose $\epsilon = 0$ in (3.109). Then it is clear that a sequence $\{z_j\}_{j \geq 1}$, with $|z_j| \geq \lambda$, $j \geq 1$, can be chosen such that one of the roots of the polynomial

$$|sI_n - A - \sum_{j=1}^{M} B_{j-1} z_i^{1-j} P(z_i)^{-1} C|, \quad i \geq 1$$

approaches a finite point $s = i\omega_0$ on the imaginary axis of the complex plane as $i \to +\infty$. Further, suppose, without loss of generality, that the sequence $\{z_j^{-1}\}$ converges to a point $z = z_0$ where the assumption (iii) guarantees that $z_0 \neq \gamma_i$, $1 \leq i \leq q$, and there are now two possibilities (a) $z_0 = 0$ and (b) $z_0 \neq 0$. If $z_0 \neq 0$, then the controllability and observability conditions imply that $||(z_0^{-1} I - L_\alpha)^{-1}|| \to +\infty$ which violates the requirement of stability along the pass. Alternatively, if $z_0 = 0$ then A must have at least one pair of purely imaginary eigenvalues and hence, using the controllability and observability assumptions, $||L_\alpha|| \to +\infty$ as $\alpha \to +\infty$ which violates (3.60) of theorem 3.3.2 for stability along the pass. ∎

Setting $M = 1$ in the above theorem now gives the following result for the process of example 2.3.4.

Corollary 3.3.8: The differential unit memory case - Suppose that the pair $\{C,A\}$ is observable and the pair $\{A,B_0\}$ is controllable. Then the extended linear repetitive process $S(E_\alpha, W_\alpha, L_\alpha)_{\alpha \geq \alpha_0}$ generated by the model of example 2.3.4 with $\alpha \geq \alpha_0$ is stable along the pass if, and only if,

  (a) all eigenvalues of the $m \times m$ matrix D_1 lie in the open unit circle
       in the complex plane; and

  (b) there exists real numbers $\epsilon > 0$ and $r_\infty < \lambda < 1$ such that all
       eigenvalues of the $n \times n$ matrix $A + B_0 (zI_m - D_1)^{-1} C$ lie to the left
       of the line $\text{Re}\{s\} = -\epsilon$ for all choices of $|z| \geq \lambda$. ∎

The corresponding results for the discrete processes of examples 2.3.7 and 2.3.8 take the following forms

Theorem 3.3.6: The discrete non-unit memory case - Suppose that

  (i) the pair $\{C, \Phi\}$ is observable,

  (ii) the pair $\{\Phi, \sum_{j=1}^{M} \Delta_{j-1} \gamma^{j-1}\}$ is controllable at all but a finite

       number of points $\gamma_1, \gamma_2, \ldots, \gamma_q$ in the complex plane,

and

  (iii) $|z_1 I_n - \Phi - \sum_{j=1}^{M} \Delta_{j-1} \gamma_i^{j-1} P(\gamma_i^{-1})^{-1} C|$ has no roots on the unit circle in the complex plane, $1 \leq i \leq q$, where $P(\gamma)$ is defined by (3.38) with z replaced by γ.

Then the extended linear repetitive process $S(E_\alpha, W_\alpha, L_\alpha)_{\alpha \geq \alpha_0}$ generated by discrete models of the form of (2.45) in example 2.3.7 with $\alpha \geq \alpha_0$ is stable along the pass if, and only if,

(a) $r_\infty = \sup\{|z| : P(z) = 0\} < 1$, (3.111)

and

(b) there exists real numbers $\epsilon > 0$ and $r_\infty < \lambda < 1$ such that

$$|z_1 I_n - \Phi - \sum_{j=1}^{M} \Delta_{j-1} z^{1-j} P(z)^{-1} C| \neq 0$$ (3.112)

for all complex numbers z_1, z satisfying $|z_1| \geq 1 - \epsilon$ and $|z| \geq \lambda$. ∎

Corollary 3.3.9: The discrete unit memory case - Suppose that the pair $\{C, \Phi\}$ is observable and the pair $\{\Phi, \Delta_0\}$ is controllable. Then the extended linear repetitive process $S(E_\alpha, W_\alpha, L_\alpha)_{\alpha \geq \alpha_0}$ generated by the model of example 2.3.8 with $\alpha \geq \alpha_0$ is stable along the pass if, and only if,

(a) all eigenvalues of the m×m matrix D_1 lie in the open unit circle in the complex plane; and

(b) there exists real numbers $\epsilon > 0$ and $r_\infty < \lambda < 1$ such that all eigenvalues of the n×n matrix $\Phi + \Delta_0 (z I_m - D_1)^{-1} C$ have modulus strictly less than $1 - \epsilon$ for all choices of $|z| \geq \lambda$. ∎

Note: It is assumed in this work that the controllability and observability assumptions of theorems 3.3.5 and 3.3.6 always hold.

Consider now the problem of testing the conditions of (a) and (b) of theorem 3.3.5 or 3.3.6 for a given example. Then it is immediately clear that testing (b) in either case is not a computationally feasible proposition. Further, as a step towards the development of equivalent results which are computationally feasible to test, it is convenient to introduce the following definitions at this stage.

Definition 3.3.3: The asymptotic stability polynomial, $P_a(z)$, for the process of example 2.3.3 or 2.3.7 is defined by

$$P_a(z) = |Q(z)|$$ (3.113)

where

$$Q(z) = I_m - z^{-1} D_1 - \ldots - z^{-M} D_M$$ (3.114)

and is to be regarded as a polynomial in z^{-1}. ∎

It is now a simple matter to show that (3.32)(theorem 3.1.4) for asymptotic stability in either case can be replaced by

$$|Q(z)| \neq 0, \quad \forall |z| \geq 1 \tag{3.115}$$

Definition 3.3.4: The stability along the pass polynomial, $A_p(s,z)$, for the process of example 2.3.3 is defined by

$$A_p(s,z) = |sI_n - A - \sum_{j=1}^{M} B_{j-1} z^{-j} Q(z)^{-1} C| \tag{3.116}$$

and is to be regarded as a polynomial in s with coefficients which are rational functions in z^{-1}. ∎

Definition 3.3.5: The stability along the pass polynomial, $A_p(z_1,z)$, for the process of example 2.3.7 is defined by

$$A_p(z_1,z) = |z_1 I_n - \Phi - \sum_{j=1}^{M} \Delta_{j-1} z^{-j} Q(z)^{-1} C| \tag{3.117}$$

and is to be regarded as a polynomial in z_1 with coefficients which are rational functions in z^{-1}. ∎

A simple argument now shows that (b) of theorem 3.3.5 is equivalent to the existence of real numbers $\epsilon > 0$ and $r_\infty < \lambda < 1$ such that

$$A_p(s,z) \neq 0 \tag{3.118}$$

for all complex numbers z,s satisfying $|z| \geq \lambda$ and $\text{Re}\{s\} \geq -\epsilon$. Similarly, it is easily shown that (b) of theorem 3.3.6 is equivalent to the existence of real numbers $\epsilon > 0$ and $r_\infty < \lambda < 1$ such that

$$A_p(z_1,z) \neq 0 \tag{3.119}$$

for all complex numbers z_1,z satisfying $|z_1| \geq 1 - \epsilon$ and $|z| \geq \lambda$.

The following results now provide alternative sets of conditions for stability along the pass of the processes of examples 2.3.3, 2.3.4, 2.3.7 and 2.3.8 which are computationally feasible, see chapter 4, to test. These express stability along the pass in terms of the derived conventional linear systems of section 2.4 and the 2D transfer-function matrix descriptions of section 2.5. Their effective action is to replace (b) in each case by two equivalent conditions which are computationally feasible to test. Further, the results for the discrete processes of examples 2.3.7 and 2.3.8 are again stated without proof since they follow from identical arguments to those used in establishing their differential counterparts.

Theorem 3.3.7: The differential non-unit memory case - With the assumptions of theorem 3.3.5, the extended linear repetitive process $S(E_\alpha, W_\alpha, L_\alpha)_{\alpha \geq \alpha_0}$ generated by differential models of the form of (2.29) in example 2.3.3 with $\alpha \geq \alpha_0$ is stable along the pass if, and only if,

(a) all eigenvalues of the N×N block companion matrix D, constructed from the 2D transfer-function matrix G(s,z) using (2.91) - (2.92), have modulus strictly less than unity;

(b) all eigenvalues of the matrix A have strictly negative real parts or, equivalently, the derived conventional linear system $L_D(A,B,C,D_o)$ of (2.50) is stable; and

(c) all eigenvalues of the N×N block companion frequency response matrix obtained by setting s = iω in the interpass transfer-function matrix G(s), constructed from G(s,z) using (2.90), have modulus strictly less unity for all real frequencies ω ≥ 0.

Proof: This consists of showing that (a) above is equivalent to (a) of theorem 3.3.5 and that (b) and (c) are, together, equivalent to (b) of this same result.

Consider first, therefore, (a) and note that (3.108) is equivalent to

$$\rho_D(z) = z^N |Q(z)|$$

$$= |z^M I_m - z^{M-1}D_1 - z^{M-2}D_2 - \ldots - D_M| \neq 0, \ \forall \ |z| \geq 1$$

$$(3.120)$$

Further, use of induction on M and Schur's formula, yields

$$\rho_D(z) = |zI_N - D| \qquad (3.121)$$

i.e. the characteristic polynomial of D.

To generate (b), let $|z| \rightarrow +\infty$ in (3.118) and hence all eigenvalues of A must have strictly negative real parts. Equivalently, $L_D(A,B,C,D_o)$ must be stable.

Given (a) and (b), note that

$$A_p(s,z) = \frac{|sI_n - A|}{\rho_D(z)} |zI_N - G(s)| \qquad (3.122)$$

and hence (b) of theorem 3.3.5 reduces to

$$|zI_N - G(s)| \neq 0, \ \forall \ |z| \geq \lambda, \ Re\{s\} \geq -\epsilon \qquad (3.123)$$

Setting s = iω, the necessity of (c) above follows immediately since (3.123) implies that all eigenvalues of G(iω) have modulus strictly less than λ.

Conversely, suppose that (a) - (c) above hold and consider the usual Nyquist contour in the complex plane. Further, let $z_j(s)$, $1 \leq j \leq N$, denote the jth eigenvalue of G(s) and λ_j, $1 \leq j \leq N$, the jth eigenvalue of D where $|\lambda_j| < 1$ by (a). In which case (2.91) indicates that the choice of

$$\underset{|s| \rightarrow +\infty}{\text{limit}} \ z_j(s) = \lambda_j, \ 1 \leq j \leq N \qquad (3.124)$$

incurs no loss of generality. Hence, using (a) and (c), it is possible to choose a real scalar λ in the non-empty range

$$\underset{\omega \geq 0}{\sup} \ \underset{1 \leq j \leq N}{\max} \ |z_j(i\omega)| < \lambda < 1 \qquad (3.125)$$

Now suppose that s traverses the Nyquist contour in a clockwise manner. Then it follows immediately that the locus generated by the right-hand side of (3.122) does not intersect or encircle the origin of the complex plane for any choice of $|z| \geq \lambda$. Equivalently, all eigenvalues of the matrix $A + \sum_{j=1}^{M} B_{j-1} z^{-j} \mathbb{Q}(z)^{-1} C$ have strictly negative real parts for all choices of $|z| \geq \lambda$, i.e. $A_p(s,z) \neq 0$ for all complex numbers z,s satisfying $|z| \geq \lambda$ and $\mathrm{Re}\{s\} \geq 0$.

The final step in this proof is to demonstrate the existence of a suitable $\epsilon > 0$. To accomplish this, first note again that the eigenvalues of $A + \sum_{j=1}^{M} B_{j-1} z^{-j} \mathbb{Q}(z)^{-1} C$ approach the eigenvalues of A as $|z| \rightarrow + \infty$. Hence (b) of theorem 3.3.5 can be replaced by the requirement that all eigenvalues of $A + \sum_{j=1}^{M} B_{j-1} z^{-j} \mathbb{Q}(z)^{-1} C$ have strictly negative real parts for all z lying in some compact set $\lambda \leq |z| \leq R$ with R 'large'. The existence of a suitable $\epsilon > 0$ now follows by continuity and the consequent existence of a finite covering of this set by open balls. ∎

Setting M = 1 in the above theorem now gives the following result for the process of example 2.3.4.

Corollary 3.3.10: The differential unit memory case - With the assumptions of corollary 3.3.8, the extended linear repetitive process $S(E_\alpha, W_\alpha, L_\alpha)_{\alpha \geq \alpha_0}$ generated by the model of example 2.3.4 with $\alpha \geq \alpha_0$ is stable along the pass if, and only if,

(a) all eigenvalues of the m×m matrix D_1 lie in the open unit circle in the complex plane;

(b) all eigenvalues of the matrix A have strictly negative real parts or, equivalently, $L_D(A,B,C,D_0)$ is stable; and

(c) all eigenvalues of the m×m interpass transfer-function matrix $G(s) \equiv G_1(s)$ of (2.90) with $s = i\omega$ have modulus strictly less than unity for all real frequencies $\omega \geq 0$.

∎

At this stage, consider the special case when the example under consideration is SISO. Then here (c) of corollary 3.3.10 reduces to the modulus condition

$$|G_1(i\omega)| < 1, \quad \forall \text{ real } \omega \geq 0 \qquad (3.126)$$

Equivalently, the frequency response plot generated by the interpass transfer-function $G_1(s)$, $s = i\omega$, $\forall$ real $\omega \geq 0$, must lie entirely within the unit circle in the complex plane. Suppose also that zero state initial conditions and control inputs are applied, i.e. $d_{k+1} = 0$, $U_{k+1}(t) = 0$, $0 \leq t \leq \alpha$, $k \geq 0$. Then,

after a little manipulation, the process description in this special case can be
expressed as

$$Y_k(i\omega) = G_1^k(i\omega)Y_0(i\omega), \quad k \geq 0 \tag{3.127}$$

Hence (3.126) requires that each frequency component of the initial profile is
attenuated from pass to pass. A conclusion which provides a physical interpretation
of (c) of theorem 3.3.7 in this special case.

   The corresponding results for the discrete processes of examples 2.3.7 and
2.3.8 take the following forms.

<u>Theorem 3.3.8</u>:   The discrete non-unit memory case - With the assumptions of theorem
3.3.6, the extended linear repetitive process $S(E_\alpha, W_\alpha, L_\alpha)_{\alpha \geq \alpha_0}$ generated by discrete

models of the form of (2.45) in example 2.3.7 with $\alpha \geq \alpha_0$ is stable along the pass
if, and only if,
   (a)   all eigenvalues of the N×N block companion matrix D, constructed
        from the 2D transfer-function matrix $G(z_1, z)$ using (2.103), have

        modulus strictly less than unity;
   (b)   all eigenvalues of the matrix Φ have modulus strictly less than
        unity or, equivalently, the derived conventional linear system
        $L_D(\Phi, \Delta, C, D_0)$ is stable; and
   (c)   all eigenvalues of the N×N interpass transfer-function matrix
        $G(z_1)$, constructed from $G(z_1, z)$ using (2.102), have modulus

        strictly less than unity for all frequencies z_1 satisfying $|z_1| = 1$. ∎

<u>Corollary 3.3.11</u>:   The discrete unit memory case - With the assumptions of corollary
3.3.9, the extended linear repetitive process $S(E_\alpha, W_\alpha, L_\alpha)_{\alpha \geq \alpha_0}$ generated by the model

of example 2.3.8 with $\alpha \geq \alpha_0$ is stable along the pass if, and only if,
   (a)   all eigenvalues of the m×m matrix D_1 lie in the open unit circle

        in the complex plane;
   (b)   all eigenvalues of the matrix Φ have modulus strictly less than
        unity or, equivalently, $L_D(\Phi, \Delta, C, D_0)$ is stable; and
   (c)   all eigenvalues of the m×m interpass transfer-function matrix
        $G(z_1) \equiv G_1(z_1)$ of (2.102) have modulus strictly less than unity

        for all frequencies z_1 satisfying $|z_1| = 1$.

                                                                         ∎

3.4   <u>A 2D Systems Approach</u>
   This section considers the link between repetitive process stability and BIBO
stability of 2D linear systems described by the Roesser state-space model of (2.58).
In particular, the link between stability along the pass of example 2.3.8 and BIBO

stability of (2.58) is considered. The following results summarise the essential elements of the well established stability theory for 2D linear systems (not necessarily described by the Roesser model).

In general, a 2D linear shift invariant system can be described by a convolution of the input $U(p,q)$ and the impulse response function $h(p,q)$. Here, however, it is only necessary to consider initially the special case of SISO systems described by the input/output map

$$Y(p,q) = \sum_{k=0}^{K} \sum_{\ell=0}^{L} a(k,\ell)U(p-k,q-\ell) - \sum_{i=0}^{I} \sum_{j=0}^{J} b(i,j)Y(p-i,q-j) \tag{3.128}$$

Further, (3.128) is said to be spatially causal over the quadrant $(p,q) > 0$ since $Y(p,q)$ depends only on input and output variables at points $(i,j) \leq (p,q)$.

Applying the 2D z transform to (3.128) yields a 2D transfer-function relating Y to U of the form

$$H(z_1,z_2) = \frac{A(z_1,z_2)}{B(z_1,z_2)} \tag{3.129}$$

where

$$A(z_1,z_2) = \sum_{k=0}^{K} \sum_{\ell=0}^{L} a(k,\ell)z_1^k z_2^\ell \tag{3.130}$$

and

$$B(z_1,z_2) = \sum_{i=0}^{I} \sum_{j=0}^{J} b(i,j)z_1^i z_2^j \tag{3.131}$$

and, for simplicity, it has been assumed that $b(0,0) = 1$. Further, expanding H as a power series yields

$$H(z_1,z_2) = \sum_{p=0}^{\infty} \sum_{q=0}^{\infty} h(p,q)z_1^p z_2^q \tag{3.132}$$

and (3.128) is said to be BIBO stable if, and only if,

$$\sum_{p=0}^{\infty} \sum_{q=0}^{\infty} |h(p,q)| < +\infty \tag{3.133}$$

The following standard result, stated without proof, now gives a condition for (3.133).

Theorem 3.4.1: Suppose that the two variable polynomials A and B are mutually prime and $H(z_1,z_2)$ has no nonessential singularities of the second kind (i.e. there exists no $(\hat{z}_1,\hat{z}_2)$ such that $A(\hat{z}_1,\hat{z}_2) = B(\hat{z}_1,\hat{z}_2) = 0$). Then BIBO stability of (3.128) is equivalent to

$$B(z_1,z_2) \neq 0, \quad \forall \ |z_1| \leq 1, \ |z_2| \leq 1 \tag{3.134}$$

Testing theorem 3.4.1 for a given example would clearly be a formidable task. This problem can, however, be simplified by use of the following equivalent standard result which is again stated without proof.

Theorem 3.4.2: With assumptions of theorem 3.4.1, BIBO stability of (3.128) requires that

(a) $B(z_1,0) \neq 0, \quad \forall \ |z_1| \leq 1$ (3.135)

and

(b) $B(z_1,z_2) \neq 0, \quad \forall \ |z_1| = 1, \ |z_2| \leq 1$ (3.136) ∎

Note also that (3.135) and (3.136) are interchangeable in terms of z_1 and z_2.
Further, additional simplifications of them have been derived and used, for example, to develop a 'Nyquist-like' stability test for 2D linear systems.

Note: The 2D transform, and resulting 2D transfer-function matrix, used in this section is distinct from $G(z_1,z)$ of section 2.5.

Consider now the Roesser state-space model of (2.58). Then application of the 2D z-transform yields the following 2D transfer-function matrix

$$H(z_1,z_2) = [C_1 \ C_2] \begin{bmatrix} z_1^{-1} I_{n_1} - A_1 & -A_2 \\ \\ -A_3 & z_2^{-1} I_{n_2} - A_4 \end{bmatrix}^{-1} \begin{bmatrix} B_1 \\ \\ B_2 \end{bmatrix}$$
$$+ D$$ (3.137)

Further, application of the above results to each element in turn of $H(z_1,z_2)$ immediately yields that BIBO stability, as expressed by theorem 3.4.1 or 3.4.2, is dependent on the roots of the characteristic polynomial

$$\rho(z_1,z_2) = \begin{vmatrix} I_{n_1} - z_1 A_1 & -z_1 A_2 \\ \\ -z_2 A_3 & I_{n_2} - z_2 A_4 \end{vmatrix}$$ (3.138)

Use of Schur's formula now yields
$$\rho(z_1,z_2) = |I_{n_1} - z_1 A_1||I_{n_2} - z_2 A_4 - z_1 z_2 A_3 (I_{n_1} - z_1 A_1)^{-1} A_2|$$ (3.139)

and leads to the following result.

Theorem 3.4.3: The conditions of theorem 3.4.1 or 3.4.2 are equivalent to the following:

(a) all eigenvalues of the matrix A_1 have modulus strictly less than unity;

(b) all eigenvalues of the matrix A_4 have modulus strictly less than unity; and

(c) all eigenvalues of the transfer-function matrix

$$Q(z_1^{-1}) := A_4 + A_3(z_1^{-1}I_{n_1} - A_1)^{-1}A_2 \tag{3.140}$$

with $|z_1| = 1$ lie in the open unit circle in the complex plane.

Proof: Applying (a) of theorem 3.4.2 to $\rho(z_1, z_2)$ requires that

$\rho(z_1, 0) = |I_{n_1} - z_1 A_1| \neq 0$ for $|z_1| \leq 1$ and, by interchanging the roles of z_1 and

z_2, $\rho(0, z_2) = |I_{n_2} - z_2 A_4| \neq 0$ for $|z_2| \leq 1$. Hence (a) and (b) above follow

immediately and, using (3.139), (b) of theorem 3.4.2 reduces to

$$T(z_1, z_2) := |I_{n_2} - z_2 Q(z_1^{-1})| \neq 0, \ \forall \ |z_1| = 1, \ |z_2| \leq 1 \tag{3.141}$$

i.e. all eigenvalues of $Q(z_1^{-1})$ with $|z_1| = 1$ lie in the open unit circle in the

complex plane.

Conversely, suppose that (a) - (c) above hold. Then it follows immediately
that $\rho(z_1, 0) \neq 0$ for all $|z_1| \leq 1$ and hence (3.135) is valid. Further,

$\rho(z_1, z_2) \equiv \rho(z_1, 0) T(z_1, z_2) \neq 0$ for $|z_1| = 1$, $|z_2| \leq 1$ as the eigenvalues of

$I_{n_2} - z_2 Q(z_1^{-1})$ are non-zero in this domain and the proof is complete.

∎

Theorem 3.4.3 is in its own right an alternative to theorems 3.4.1 or 3.4.2 for
BIBO stability of the Roesser model. In this context, (a) and (b) are equivalent
necessary conditions and hence one of them could be dispensed with. They are
retained here, however, since the primary purpose is to establish an equivalence
with stability along the pass of the discrete unity memory linear repetitive process
of example 2.3.8. This is contained in the following result.

Theorem 3.4.4: Regard the model of example 2.3.8 as a 2D system described by the
Roesser model and suppose that the corresponding 2D transfer-function matrix of
(3.137) has no nonessential singularities of the second kind. Then the extended
linear repetitive process $S(E_\alpha, W_\alpha, L_\alpha)_{\alpha \geq \alpha_0}$ generated by this model with $\alpha \geq \alpha_0$ is

stable along the pass if, and only if, it is BIBO stable in the sense of theorem
3.4.3.

Proof: This, in effect, consists of showing that the conditions of theorem 3.4.3
and corollary 3.3.11 are equivalent.

Suppose first, therefore, that theorem 3.4.2 is applied. Then this requires
the following conditions for BIBO stability which are precisely those of corollary
3.3.11 for stability along the pass.

(a) all eigenvalues of the matrix Φ have modulus strictly less than unity;

(b) all eigenvalues of the matrix D_1 have modulus strictly less than unity; and

(c) all eigenvalues of

$$G_1(z_1^{-1}) := C(z_1^{-1} I_n - \Phi)^{-1}\Delta_0 + D_1 \tag{3.142}$$

with $|z_1| = 1$ lie in the open unit circle in the complex plane.

Conversely, suppose that (a) - (c) above hold. Then the proof that these imply stability along the pass is identical to that of corollary 3.3.11 and is hence omitted. ∎

The major conclusion to be drawn from this result is that any one of numerous tests for BIBO stability of 2D linear systems described by the Roesser model can be applied to the linear repetitive process of example 2.3.8. In particular, regard the model of this example as a 2D linear system described by the Roesser model with 2D transfer-function matrix (3.137) which is assumed to have no nonessential singularities of the second kind. Further, define the following two variable polynomial in terms of this transfer-function matrix

$$\rho(z_1,z_2) = \begin{vmatrix} I_n - z_1\Phi & -z_1\Delta_0 \\ -z_2 C & I_m - z_2 D_1 \end{vmatrix} \tag{3.143}$$

Then, for example, the following 2D stability tests are applicable to example 2.3.8.

Corollary 3.4.4: The extended linear repetitive process $S(E_\alpha, W_\alpha, L_\alpha)_{\alpha \geq \alpha_0}$ generated by the model of example 2.3.8 with $\alpha \geq \alpha_0$ is stable along the pass if, and only if,

$$\rho(z_1,z_2) \neq 0, \quad \forall \; |z_1| \leq 1, \; |z_2| \leq 1 \tag{3.144}$$
∎

Corollary 3.4.5: The extended linear repetitive process $S(E_\alpha, W_\alpha, L_\alpha)_{\alpha \geq \alpha_0}$ generated by the model of example 2.3.8 with $\alpha \geq \alpha_0$ is stable along the pass if, and only if,

$$\rho(z_1,0) \neq 0, \quad \forall \; |z_1| \leq 1 \tag{3.145}$$

and

$$\rho(z_1,z_2) \neq 0, \quad \forall \; |z_1| = 1, \; |z_2| \leq 1 \tag{3.146}$$
∎

Corollary 3.4.6: The extended linear repetitive process $S(E_\alpha, W_\alpha, L_\alpha)_{\alpha \geq \alpha_0}$ generated by the model of example 2.3.8 with $\alpha \geq \alpha_0$ is stable along the pass if, and only if there exists a,b with $|a| < 1$, $|b| = 1$, such that

(a) $\quad \rho(a, z_2) \neq 0, \quad \forall \quad |z_2| \leq 1$ $\hfill$ (3.147)

(b) $\quad \rho(z_1, b) \neq 0, \quad \forall \quad |z_1| \leq 1$ $\hfill$ (3.148)

and

(c) $\quad \rho(z_1, z_2) \neq 0, \quad \forall \quad |z_1| = 1, \; |z_2| = 1$ $\hfill$ (3.149)

These and other conditions arising from theorem 3.4.4 are considered again in chapter 4 where the general subject is the development of computationally feasible stability tests.

Notes and References

The results up to and including theorem 3.3.5 are based on the original work of Owens (1977) which was extended by Rogers (1987). For the necessary functional analysis see, for example, Taylor (1958). Theorems 3.3.7 and 3.3.8, which express stability along the pass in terms of the 2D transfer-function matrix, are due to Rogers and Owens (1989a,1990a) Section 3.4 has evolved from the work of Boland and Owens (1980), where the 2D z-transform and theorem 3.4.1 are due to Shanks, Treitel and Justice (1972), theorem 3.4.2 is due to Huang (1972), and corollary 3.4.6 is from Strintzis (1977).

CHAPTER 4

GRAPHICAL AND ALGEBRAIC STABILITY TESTS

This chapter develops computationally feasible tests for stability along the pass of the differential and discrete processes of examples 2.3.3 and 2.3.7 from theorems 3.3.7 and 3.3.8 respectively. The end product is two systematic test procedures in each case. These procedures are also compared from an applications standpoint with particular emphasis on CAD (Computer Aided Design) aspects. The equivalence developed in section 3.4 between BIBO stability of 2D linear systems described by the Roesser model and stability along the pass of example 2.3.8 (the discrete unit memory case) is considered in depth from a stability analysis/controller design standpoint. Finally, the application of results from the stability theory of delay differential systems to example 2.3.4 (the differential unit memory case) is analysed from the same standpoint.

4.1 Asymptotic Stability

Consider the extended linear repetitive process $S(E_\alpha, W_\alpha, L_\alpha)_{\alpha \geq \alpha_0}$ generated by the model of example 2.3.3 or 2.3.7. Then, by (a) of theorem 3.3.7 or 3.3.8 as appropriate, this process is asymptotically stable for all $\alpha \geq \alpha_0$ if, and only if, all eigenvalues of the N×N block companion matrix

$$D = \begin{bmatrix} 0 & I_m & & 0 \\ 0 & 0 & & I_m \\ D_M & D_2 & & D_1 \end{bmatrix} \tag{4.1}$$

have modulus strictly less than unity. Note also that this condition is necessary for stability along the pass and hence no further tests are required if it does not hold. Further, suppose that the elements of D_j, $1 \leq j \leq M$, are known numerically. Then the obvious CAD orientated test is to simply compute the eigenvalues of D and display them relative to the unit circle in the complex plane.

Alternatively, write
$$\rho_D(z) = |zI_N - D|$$
$$= a_N z^N + a_{N-1} z^{N-1} + \ldots + a_1 z + a_0 \tag{4.2}$$
where the coefficients are real scalars with $a_N = 1$. Then (a) of theorem 3.3.7 or 3.3.8 is equivalent to

$$\rho_D(z) \neq 0, \quad \forall \; |z| \geq 1 \tag{4.3}$$

i.e. all roots of $\rho_D(z)$ must lie in the open unit circle in the complex plane. This is just the standard stability result from discrete conventional linear systems theory and hence it can be tested by applying any one of numerous well established tests which avoid the need to compute the roots of $\rho_D(z)$. At this stage, however, only the Schur-Cohn matrix test is considered since, as shown in later sections, it can also be used to develop tests for other conditions of theorems 3.3.7 and 3.3.8. An alternative, the so-called Jury/Marden table test, will be introduced in section 4.3 where it will play a particular role in the development of one possible test for (c) of theorem 3.3.8.

The Schur-Cohn matrix test, see the cited reference for the relevant background, converts the problem of determining whether or not all roots of $\rho_D(z)$ lie inside the unit circle of the complex plane to one of determining whether or not a real symmetric matrix constructed from its coefficients is positive definite. In particular, suppose that the N×N symmetric matrix $H = \{h_{ij}\}$ where

$$h_{ij} = \sum_{k=1}^{i} (a_{N-i+k}a_{N-j+k} - a_{i-k}a_{j-k}), \; i \leq j \tag{4.4}$$

is constructed from the coefficients of $\rho_D(z)$. Then it can be shown that (a) of theorem 3.3.7 or 3.3.8 holds if, and only if, H of (4.4) is positive definite. Further, this new condition can be tested by any one of numerous equivalent criteria. Hence, for example, the following result constitutes a test for (a) of theorem 3.3.7 or 3.3.8.

Lemma 4.1.1: Consider the extended linear repetitive process $S(E_\alpha, W_\alpha, L_\alpha)_{\alpha \geq \alpha_0}$ generated by the model of example 2.3.3 or 2.3.7 with $\alpha \geq \alpha_0$. Then (a) of theorem 3.3.7 or 3.3.8, as appropriate, for stability along the pass holds if, and only if, all principal minors of the Schur-Cohn matrix (4.4) are positive. ∎

Application of the test of lemma 4.1.1, or any other one based on $\rho_D(z)$ written in the form (4.2), is straightforward, given the coefficients. Clearly, however, obtaining these is not a particularly feasible proposition from a CAD standpoint. Further, these tests do not provide easily used measures of relative stability and/or performance indicators. Hence the major remit of such tests is clearly in low order synthesis type problems where some, or all, of the elements of the matrices D_j, $1 \leq j \leq M$, are design parameters.

4.2 Stability Along The Pass - The Differential Case

Suppose that condition (a) of theorem 3.3.7 for stability along the pass of the extended linear repetitive process generated by the model of example 2.3.3 with

$\alpha \geq \alpha_0$ holds and consider the condition listed under (b). Note also that this condition is necessary for stability along the pass and hence no further tests are required if it does not hold. Further, suppose that the elements of A are known numerically. Then the obvious CAD orientated test is, as in the case of (a), to compute the eigenvalues of A and display them relative to the open left-half of the complex plane.

Alternatively, write

$$\rho_A(s) = |sI_n - A|$$
$$= a_n s^n + a_{n-1} s^{n-1} + \ldots + a_1 s + a_0 \tag{4.5}$$

where the coefficients are real scalars with $a_n = 1$. Then (b) of theorem 3.3.7 is equivalent to

$$\rho_A(s) \neq 0, \quad \forall \text{ Re}\{s\} \geq 0 \tag{4.6}$$

i.e. all roots of $\rho_A(s)$ lie in the open left-half of the complex plane. Further, (4.6) can be tested without computing the roots of $\rho_A(s)$ by employing the classical Routh array.

Application of the Routh array test to (4.6) is straightforward, given the coefficients in (4.5). Clearly, however, obtaining these coefficients is not a particularly feasible proposition from a CAD standpoint. Further, this test does not provide easily used measures of relative stability and/or performance indicators. Hence, as with the tests based on $\rho_D(z)$ written in the form of (4.2) for (a) of theorem 3.3.7, the major remit of this test is clearly in low order synthesis type problems where some, or all, of the elements of A are design parameters.

Suppose now that (a) and (b) of theorem 3.3.7 hold and then the particular example under consideration is stable along the pass if, and only if, condition (c) holds. Further, consider again the interpass transfer-function matrix G(s) of (2.90), i.e.

$$G(s) = \begin{bmatrix} 0 & I_m & 0 \\ 0 & 0 & I_m \\ G_M(s) & G_2(s) & G_1(s) \end{bmatrix} \tag{4.7}$$

where

$$G_j(s) = C(sI_n - A)^{-1} B_{j-1} + D_j, \quad 1 \leq j \leq M \tag{4.8}$$

and set $s = i\omega$. Then it follows immediately that this condition is equivalent to the requirement that the continuous curves, or characteristic loci, generated by the eigenvalues $z_j(i\omega)$, $1 \leq j \leq N$, of (4.7)-(4.8) lie entirely within the unit circle in the complex plane for all real $\omega \geq 0$. Consequently testing of this condition reduces to the evaluation and representation of these loci relative to the unit circle in the complex plane. A task which, with the additional simple operation of superimposing the unit circle onto the resulting plots, can be undertaken using standard CAD software for the derived system $L_D(A,B,C,D_0)$.

Note: The $z_j(i\omega)$, $1 \leq j \leq N$, are termed the repetitive system characteristic loci to distinguish them from those associated with the transfer-function matrix, $G_0(s)$, of $L_D(A,B,C,D_0)$.

To develop an alternative test for (c) of theorem 3.3.7 to that given above, first write $\rho(z,s) = 0$ where

$$\rho(z,s) := |zI_N - G(s)| \tag{4.9}$$

in the form

$$z^N + c_{N-1}(s)z^{N-1} + \ldots + c_1(s)z + c_0(s) = 0 \tag{4.10}$$

with coefficients $c_0(s)$, $c_1(s)$, $\ldots$, $c_{N-1}(s)$ which are rational functions in s. Further, let $a_N(s)$ denote the least common denominator of the $c_j(s)$, $j = 0,1,\ldots,N-1$. Then (4.10) can be written as

$$a_N(s)z^N + a_{N-1}(s)z^{N-1} + \ldots + a_1(s)z + a_0(s) = 0 \tag{4.11}$$

where the coefficients $a_0(s)$, $a_1(s)$, $\ldots$, $a_N(s)$ are real polynomials in s. In which case it follows immediately that (c) of theorem 3.3.7 holds if, and only if, the, assumed irreducible, polynomial

$$\rho(z) = a_N(s)z^N + a_{N-1}(s)z^{N-1} + \ldots + a_1(s)z + a_0(s) \tag{4.12}$$

satisfies

$$\rho(z) \neq 0, \ \forall s: \ \text{Re}\{s\} = 0, \ |z| \geq 1 \tag{4.13}$$

In order to develop a test for (4.13), first suppose that the N×N Schur-Cohn matrix $H = \{h_{ij}\}$ is constructed from its coefficients where

$$h_{ij} = \sum_{k=1}^{i} (a_{N-i+k}\bar{a}_{N-j+k} - \bar{a}_{i-k}a_{j-k}), \ i \leq j$$

$$h_{ij} = \bar{h}_{ji}, \ i > j \tag{4.14}$$

Then each element in H is a polynomial in s and/or its complex conjugate $\bar{s}$ and it can be shown that (4.13) holds if, and only if, the Hermitian polynomial matrix $H(s) \equiv H$ is positive definite for all $s:\text{Re}\{s\} = 0$. Equivalently, for all constant complex vectors $U \neq 0$ of a unitary N-dimensional vector space,

$$U^* H(s)U > 0, \ \forall s:\text{Re}\{s\} = 0 \tag{4.15}$$

where * denotes the complex conjugate transpose.

A Hermitian polynomial matrix H(s) satisfying (4.15) is said to be axis positive and the following result gives necessary and sufficient conditions for the existence of this property.

<u>Theorem 4.2.1</u>: The N×N Hermitian polynomial matrix H(s) is axis positive if, and only if,

(a) $H(0) > 0$ (4.16)

and

(b) $|H(s)| > 0, \forall s: \text{Re}\{s\} = 0$ (4.17)

<u>Proof:</u> If $H(s) > 0$, i.e. (4.15) holds, then clearly (4.16) and (4.17) hold. Conversely, suppose that (4.16) and (4.17) hold. Then, using (4.17), $|H(i\omega)| \neq 0, \forall \omega$, i.e. the eigenvalues $\lambda_k(\omega)$, $1 \leq k \leq N$, of $H(i\omega)$ are non-zero for all ω. Further, the $\lambda_k(\omega)$ are real continuous functions of ω which are positive at $\omega = 0$ by (4.16) and hence positive $\forall \omega$.   ∎

The first condition of theorem 4.2.1 requires that the real symmetric matrix $H(0)$ is positive definite. Hence it can be tested by applying any one of numerous equivalent tests. For example, this condition holds if, and only if, all principal minors of $H(0)$ are positive.

To develop a test for (b) of theorem 4.2.1, set $s = i\omega$ and note that the determinant of a Hermitian matrix is real. Consequently $q(\omega^2) := |H(i\omega)|$ must be a real polynomial of the form

$$q(\omega^2) = \sum_{k=0}^{r} q_{2k}\omega^{2k}$$ (4.18)

and it follows immediately that (4.17) holds if, and only if, $q(\omega^2)$ has the so-called positive realness property

$$q(\omega^2) > 0, \forall \omega \geq 0$$ (4.19)

Further, (4.19) can be expressed in terms of the roots of $q(\omega^2)$ by means of the following easily proven result.

<u>Lemma 4.2.1</u>: The polynomial $q(\omega^2)$ satisfies (4.19) if, and only if, it has no positive real roots and $q(\omega^2) > 0$ for some $\omega \geq 0$.   ∎

<u>Note 1</u>: In the trivial case of $r = 0$, lemma 4.2.1 reduces to $q(\omega^2) \equiv q_0 > 0$.

<u>Note 2</u>: $q(\omega^2) > 0$ for some $\omega \geq 0$ can be replaced by either $q_0 > 0$ or $q_{2r} > 0$ [i.e. $q(+\infty) > 0$] which is easy to test. In particular, (4.19) is violated if $q_0 < 0$ and no further tests are required.

At this stage, the classical Descartes rule of sign yields the following preliminary results concerning the positive realness property (4.19).

<u>Lemma 4.2.2</u>: With $q_0 > 0$, a necessary condition for $q(\omega^2)$ to have no positive real roots is that there is an even number of changes of sign in the coefficients q_{2k} when arranged in decending order.   ■

If the number of sign variations is zero, i.e. all coefficients are positive, the above result is sufficient and is a special case of the following lemma.

<u>Lemma 4.2.3</u>: A sufficient condition for $q(\omega^2)$ to have no positive real roots is that the coefficients q_{2k} satisfy

$$q_0 > 0, \; q_{2k} \geq 0, \; 1 \leq k \leq r \qquad\qquad (4.20)$$

To test lemma 4.2.1 in the general case, requires a means of determining the location of the real roots (if any) of the real polynomial $q(\omega^2)$. This is a well researched problem and numerous solutions exist one of which, for example, uses the concept of a matrix inner. Here, however, only the test detailed below is used since it is known to be computationally less expensive to implement for numerical examples.

First note that $q(\omega^2)$ of (4.18) is a real even order polynomial and therefore has $2r$ roots symmetrically distributed with respect to both the real and imaginary axes of the complex plane. Suppose also that the polynomial

$$q(i\omega) = \sum_{k=0}^{r} (-1)^k q_{2k}\omega^{2k} \qquad\qquad (4.21)$$

is constructed from (4.18). Then, since (4.21) represents an anti-clockwise rotation of 90°, the symmetry discussed above is preserved but real roots, if any, of $q(\omega^2)$ become purely imaginary roots of $q(i\omega)$. Consequently if $q(i\omega)$ has r roots with positive real parts then $q(\omega^2)$ has no roots with positive real parts and hence lemma 4.2.1 holds under the assumption that $q_0 > 0$.

To test this new condition, replace ω^2 by ω and form the so-called, see the cited reference for the necessary background, modified Routh array

ROW							
1	ω^{2r}	$(-1)^r q_{2r}$	$(-1)^{r-1}q_{2r-2}$	· · ·	· · ·	$-q_2$	q_0
2	ω^{2r-1}	$(-1)^r rq_{2r}$	$(-1)^{r-1}(r-1)q_{2r-2}$	·	· · ·	$-q_2$	
.	.						
.	.						
.	.						
$2r+1$	ω^0	q_0					

$$\qquad\qquad (4.22)$$

for (4.21) where

(i)   the entries in row 2 are given by the coefficients of the derivative of $q(i\omega)$; and

(ii) the entries in row j, $3 \le j \le r + 1$ are constructed as for the standard Routh array.

Now let $\text{Var}[(-1)^r q_{2r},(-1)^r r q_{2r},\ldots,q_0]$ denote the number of changes of sign in the sequence $[(-1)^r q_{2r},(-1)^{r-1} r q_{2r},\ldots,q_0]$. Then Routh's result can be invoked to show that $q(i\omega)$ must have

$$Q = \text{Var}[(-1)^r q_{2r}, (-1)^r r q_{2r},\ldots,q_0] \qquad (4.23)$$

roots with positive real parts in this case. Hence it follows immediately that $q(\omega^2)$ has no positive real roots if, and only if, there are r changes of sign in the first column of the associated modified Routh array (4.22), i.e.

$$r = \text{Var}[(-1)^r q_{2r},(-1)^r r q_{2r},\ldots,q_0] \qquad (4.24)$$

As an example of the use of (4.22)-(4.24), consider the case when

$$q(\omega^2) = \omega^8 - 3\omega^6 + 2\omega^4 + \omega^2 + 1 \qquad (4.25)$$

Then

$$q(i\omega) = \omega^8 + 3\omega^6 + 2\omega^4 - \omega^2 + 1 \qquad (4.26)$$

and the array of (4.22) in this case is

ROW					
ω^8	1	3	2	-1	1
ω^7	4	9	4	-1	0
ω^6	0.75	1	-0.75	1	0
ω^5	3.66	8	-6.33	0	
ω^4	-0.64	0.545	1	0	
ω^3	11.12	-0.61	0		
ω^2	0.51	1			
ω^1	-22.4	0			
ω^0	1				

$$(4.27)$$

Here

$$\text{Var}[1,4,0.75,3.66,-0.64,11.12,0.51,-22.4,1] = 4 \qquad (4.28)$$

and hence, since r = 4, $q(\omega^2)$ of (4.25) has no positive real roots. Finally, note that these tests are not suitable for CAD implementation. Hence their major remit is clearly in low order synthesis problems where some, or all, of the elements in $G_j(s)$, $1 \le j \le M$, contain design parameters.

At this stage, two alternative tests have been developed for each of the three conditions of theorem 3.3.7. Further, it is clear that these should be tested in the order of (a) followed by (b) followed by (c) with termination if the one just tested does not hold. Suppose also that the first test is used in each case. Then the following steps represent an eigenvalue, or graphically based, systematic procedure for testing theorem 3.3.7.

STEP 1: Test the necessary condition of (a) by computing the eigenvalues of the matrix D of (4.1) and displaying them relative to the unit circle in the complex plane. Stop if this condition does not hold.

STEP 2: Test (b) by computing the eigenvalues of the matrix A and displaying them relative to the open left-half of the complex plane. Stop if this necessary condition does not hold.

STEP 3: Compute the repetitive system characteristic loci generated by the eigenvalues of the interpass transfer-function matrix $G(s)$ of (4.7), $s = i\omega$, for all real $\omega \geq 0$ and display them relative to the unit circle in the complex plane. The stability along the pass characteristics of the particular example under consideration now follow immediately on visual inspection of the resulting plots.

Using the above systematic procedure, therefore, (a) - (c) of theorem 3.3.7 can be tested using tests developed from the 2D transfer-function matrix $G(s,z)$. These tests are suitable for inclusion in a CAD package and are, in effect, well known tests from conventional linear systems theory. To illustrate the use of this procedure, consider the unit memory process described, under suitable choice of current pass state variables, by

$$\dot{X}_{k+1}(t) = \begin{bmatrix} 0 & 1 & 0 \\ 0 & 0 & 1 \\ -24 & -26 & -9 \end{bmatrix} X_{k+1}(t) + \begin{bmatrix} 2 & 0 & 0 \\ 0 & 3 & 0 \\ 0 & 0 & 4 \end{bmatrix} U_{k+1}(t)$$

$$+ \begin{bmatrix} 1 & 0 & 0 \\ 0 & 1 & 0 \\ 0 & 0 & 1 \end{bmatrix} Y_k(t)$$

$$Y_{k+1}(t) = \begin{bmatrix} 1 & 0 & 0 \\ 0 & 1 & 0 \\ 0 & 0 & 1 \end{bmatrix} X_{k+1}(t)$$

$$0 \leq t \leq \alpha, \quad X_{k+1}(0) = 0, \quad k \geq 0 \tag{4.29}$$

STEP 1: This step is redundant here since the matrix $D_1 = 0$.

STEP 2:   A simple calculation yields the eigenvalues of the matrix A as
$\lambda_1 = -2$, $\lambda_2 = -3$, $\lambda_3 = -4$. Hence (b) of theorem 3.3.7 holds.

STEP 3:   A simple calculation yields that the eigenvalues of $G(s) \equiv G_1(s)$ are given
by

$$z_1(s) = \frac{1}{s+2} , \quad z_2(s) = \frac{1}{s+3} , \quad z_3(s) = \frac{1}{s+4} \qquad (4.30)$$

The repetitive system characteristic loci generated by the elements of (4.30), s =
$i\omega$, $\forall$ real $\omega \geq 0$, are shown relative to the unit circle in the plots of Figure 4.1.
Hence (4.29) is stable along the pass.

Suppose now that each condition in theorem 3.3.7 is tested using the second of
the tests developed for it earlier in this section. Then the following steps
represent an algebraic, or root clustering, based systematic procedure for testing
this result. This serves as an alternative to the eigenvalue based procedure
detailed above.

STEP 1:   Test the necessary condition of (a) by applying an appropriate test from
the stability theory of discrete conventional linear systems. For example, the
Schur-Cohn matrix test, as defined by (4.4), could be used. Stop if this condition
does not hold.

STEP 2:   Test (b) by applying the Routh array to the characteristic polynomial,
$\rho_A(s)$ of (4.6), of the derived system $L_D[A,B,C,D_o]$. Stop if this necessary

condition does not hold.

STEP 3:   Construct $\rho(z)$ of (4.12) and hence the Schur-Cohn matrix H(s) of (4.14).
Test if H(0) is positive definite, and stop if this is not the case.

STEP 4:   Construct $q(\omega^2) = |H(i\omega)|$, and stop if lemma 4.2.2 does not hold for this
polynomial. Then apply lemma 4.2.3 and stop if this condition holds, since the
example under consideration is stable along the pass. If, however, it does not
hold, the final step is to test the modified Routh array condition of (4.24).

As noted when each of them was considered separately, the tests employed in the
above procedure are not suitable for CAD implementation. Hence the major remit of
this procedure is clearly in low order synthesis problems where some, or all, of the
matrices of the example under consideration contain design parameters. To
illustrate its application in such a case, consider the following SISO unit memory
example where a_1 and a_2 are positive real scalars:

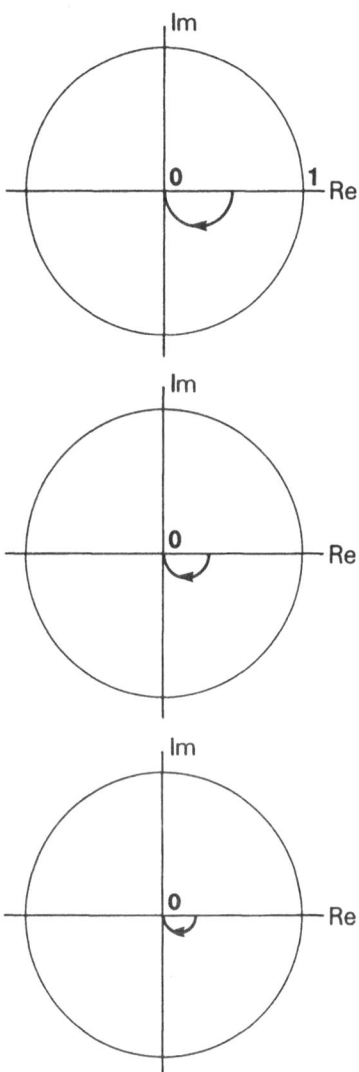

FIGURE 4.1

$$\dot{X}_{k+1}(t) = \begin{bmatrix} 0 & 1 & 0 \\ 0 & 0 & 1 \\ -1 & -3 & -3 \end{bmatrix} X_{k+1}(t) + \begin{bmatrix} 0 \\ 0 \\ 1 \end{bmatrix} U_{k+1}(t) + \begin{bmatrix} 0 \\ 0 \\ 1 \end{bmatrix} Y_k(t)$$

$$Y_{k+1}(t) = [a_1 \ a_2 \ 0]X_{k+1}(t)$$

$$0 \le t \le \alpha, \quad X_{k+1}(0) = 0, \quad k \ge 0 \tag{4.31}$$

STEP 1:   This step is redundant here since $D_1 = 0$.

STEP 2:   The characteristic polynomial of $L_D(A,B,C,D_o)$ is given by

$$\rho_A(s) = s^3 + 3s^2 + 3s + 1 \tag{4.32}$$

Application of the Routh array now shows that $L_D(A,B,C,D_o)$ is stable and hence (b)
of theorem 3.3.7 holds.

STEP 3: $$\rho(z) = (s^3 + 3s^2 + 3s + 1)z - (a_2 s + a_1) \tag{4.33}$$

$$H(s) = (s^3 + 3s^2 + 3s + 1)(\bar{s}^3 + 3\bar{s}^2 + 3\bar{s} + 1)$$
$$\quad - (a_2 s + a_1)(a_2 \bar{s} + a_1) \tag{4.34}$$

$$H(0) = 1 - a_1^2 \tag{4.35}$$

and $H(0) > 0$ if, and only if,

$$a_1 < 1 \tag{4.36}$$

STEP 4: $$q(\omega^2) = \omega^6 + 3\omega^4 + (3 - a_2^2)\omega^2 + (1 - a_1^2) \tag{4.37}$$

Further lemma 4.2.2 holds for any choice of a_2 and $a_1 < 1$. Using lemma 4.2.3, the
following constraints on a_1 and a_2 are a sufficient condition for stability along
the pass.

$$a_2 < \sqrt{3}, \quad a_1 < 1 \tag{4.38}$$

If the constraint on a_2 in (4.38) is violated, then (4.24) must be tested and two
options exist. The first of these is to construct the array with a_1 and a_2
arbitrary and then attempt to obtain conditions on a_1 and a_2 which give three
changes of sign in the first column and hence stability along the pass.
Alternatively, if particular values of a_1 and a_2 are given, construct the array and
count the number of changes of sign in the first column. As an example of this
second option, (4.39) below shows the array for the choice of $a_1 = \dfrac{1}{\sqrt{2}}$ and $a_2 = 2\sqrt{3}$.

ROW					
1	ω^6	-1	3	9	0.5
2	ω^5	-3	6	9	
3	ω^4	1	6	0.5	
4	ω^3	24	10.5		
5	ω^2	5.56	0.5		
6	ω^1	8.34			
7	ω^0	0.5			

$$(4.39)$$

Here there is only one change of sign in the first column and hence this particular case is unstable along the pass.

The first of the two systematic test procedures developed in this section uses, in effect, 'Nyquist like' tests from the stability analysis of $L_D(A,B,C,D_0)$. In this latter context, a major advantage of such tests is that they provide easily used relative stability and/or performance indicators such as gain and phase margins. These are extensively used in controller design (particularly in the SISO case) and it would obviously be desirable to have similar measures available in the repetitive systems case.

Extensive studies on a number of industrial examples have concluded that appropriately defined relative stability and/or performance indicators in the spirit of gain and phase margins should have a constructive role to play in controller design for repetitive processes. These studies have also shown, however, that computable information concerning the rate of approach of the output sequence to the limit profile (see definition 3.1.3 and theorem 3.1.2) is at least of equal importance. Further, it is clear that such information is not available from these 'Nyquist like' tests. The only option being to inspect the result of a closed-loop simulation study with the consequent prospect of a heavy computational load.

The problems discussed above are considered again in section 4.5 and in the next chapter where alternative simulation-based tests for stability along the pass are developed from suitably well behaved plant step response data which is assumed to be available. This leads to sufficient, but not necessary, stability tests which produce, at no extra cost, computable information concerning the rate of approach to the limit profile in one special case of major practical interest. Finally, the use of this information in the formulation of controller design algorithms is considered in chapter 6.

4.3 Stability Along the Pass - The Discrete Case

This section considers the testing of theorem 3.3.8 for stability along the pass of the extended linear repetitive process $S(E_\alpha, W_\alpha, L_\alpha)_{\alpha \geq \alpha_0}$ generated by the model of example 2.3.7. Here, as in the case of theorem 3.3.7 for the differential process, the first step is to test the condition listed under (a). This can be undertaken using either of the tests developed in section 4.1. As another alternative to these, the so-called Jury/Marden table test is considered below since, see later in this section, it plays a particular role in the development of one possible test for the condition listed under (c).

The Jury/Marden table test, see again the cited reference for the relevant background theory, determines the location of the roots of $\rho_D(z)$ of (4.2) relative to the unit circle using a 'Routh like' array expressed in terms of the determinants of 2×2 matrices and proceeds as follows. First construct the so-called conjugate polynomial of $\rho_D(z)$ as

$$\rho_D^*(z) = z^N \rho_D(z^{-1}) = \sum_{k=0}^{N} a_k z^{N-k} \tag{4.40}$$

and then generate the sequence of polynomials

$$g_j(z) = \sum_{k=0}^{N-j} a_k^{(j)} z^k \tag{4.41}$$

where

$$g_0(z) = \rho_D(z) \tag{4.42}$$

and

$$g_{j+1}(z) = a_0^{(j)} g_j(z) - a_{N-j}^{(j)} g_j^*(z)$$

$$j = 0, 1, \ldots, N - 1 \tag{4.43}$$

This yields the following recursive relationship for the coefficients of g_{j+1},

$$j = 0, 1, \ldots, N - 1$$
$$a_k^{(j+1)} = a_0^{(j)} a_k^{(j)} - a_{N-j}^{(j)} a_{N-j-k}^{(j)} \tag{4.44}$$

Further, denote the constant term, $a_0^{(j)}$, of $g_{j+1}(z)$ by δ_{j+1}, i.e.

$$\delta_{j+1} = (a_0^{(j)})^2 - (a_{N-j}^{(j)})^2 \tag{4.45}$$

and define the scalars P_k, $1 \leq k \leq N$, as

$$P_k = \delta_1 \delta_2 \cdots \delta_k \tag{4.46}$$

Suppose also that none of the P_k are zero and let V of them be negative. Then it can be shown that the following result holds which serves as an alternative to lemma 4.1.1 for (a) of theorem 3.3.8 (or (a) of theorem 3.3.7)).

Lemma 4.3.1: Consider the extended linear repetitive process $S(E_\alpha, W_\alpha, L_\alpha)_{\alpha \geq \alpha_0}$ generated by the model of example 2.3.7 with $\alpha \geq \alpha_0$. Then (a) of theorem 3.3.8 for stability along the pass holds if, and only if, $V = N$. ∎

Given a particular $\rho_D(z)$, the following steps represent a systematic procedure for computing V.

STEP 1: Construct the following so-called Jury/Marden table from the coefficients $a_0, a_1, \ldots, a_N$.

ROW	z^0	z^1	z^2	$\ldots$	z^k	$\ldots$	z^{N-1}	z^N
1	a_0	a_1	a_2	$\ldots$	a_{N-k}	$\ldots$	a_{N-1}	a_N
2	a_N	a_{N-1}	a_{N-2}	$\ldots$	a_k	$\ldots$	a_1	a_0
3	$\delta_1 = b_0$	b_1	b_2	$\ldots$	$..$	$\ldots$	b_{N-1}	
4	b_{N-1}	b_{N-2}	b_{N-3}	$\ldots$	$..$	$\ldots$	b_0	
5	$\delta_2 = c_0$	c_1	c_2	$\ldots$	$..$	c_{N-2}		
6	c_{N-2}	c_{N-3}	c_{N-4}	$\ldots$	$..$	c_0		
.	.	.						
.	.	.						
2N-1	$\delta_{N-1} = r_0$	r_1						
2N	r_1	r_0						
2N+1	$\delta_N = t_0$							(4.47)

Here the entries in row 2k+2 consist of the entries in row 2k+1 written in reverse order k = 0,1,2,...

and

$$
b_k = \begin{vmatrix} a_0 & a_{N-k} \\ a_N & a_k \end{vmatrix}, \qquad c_k = \begin{vmatrix} b_0 & b_{N-1-k} \\ b_{N-1} & b_k \end{vmatrix}, \quad \ldots \ldots
$$

$$
t_0 = \begin{vmatrix} r_0 & r_1 \\ r_1 & r_0 \end{vmatrix} \tag{4.48}
$$

STEP 2: The numbers $\delta_1, \delta_2, \ldots, \delta_N$ are now given as the first entries in rows $3, 5 \ldots$, $2N+1$ of (4.47). It now follows immediately that all of the P_k of (4.46) are negative, and hence lemma 4.3.1 holds, if, and only if,

$$b_0 < 0, \ c_0 > 0, \ldots, \ r_0 > 0, \ t_0 > 0 \tag{4.49}$$

The following special cases may occur in applying the above table test to a given $\rho_D(z)$

(a) $P_k \neq 0$, but $g_{j+1}(z) \equiv 0$ for some $k < N$

(b) $\delta_1, \delta_2, \ldots, \delta_k \neq 0$, but

$$\delta_{k+1} = a_0^{(k+1)} = (a_0^{(k)})^2 - (a_{N-k}^{(k)})^2 = 0 \tag{4.50}$$

In either case, the modification detailed in the cited reference should be used to continue the test. Note also that it may be necessary to use this modification more than once for a particular $\rho_D(z)$.

Suppose now that (a) of theorem 3.3.8 holds and consider the condition listed under (b). Note also that this condition is necessary for stability along the pass and hence no further tests are required if it does not hold. Further, suppose that the elements of Φ are known numerically. Then the obvious CAD orientated test is to compute the eigenvalues of Φ and display them relative to the unit circle in the complex plane.

Alternatively, write

$$\rho_\Phi(z_1) = |z_1 I_n - \Phi|$$
$$= b_n z_1^n + b_{n-1} z_1^{n-1} + \ldots + b_1 z_1 + b_0 \tag{4.51}$$

where the coefficients are real scalars with $b_n = 1$. Then (b) of theorem 3.3.8 is equivalent to

$$\rho_\Phi(z_1) \neq 0, \ \forall \ |z_1| \geq 1 \tag{4.52}$$

i.e. all roots of $\rho_\Phi(z_1)$ lie in the open unit circle in the complex plane. Further, (4.52) can be tested without computing the roots of $\rho_\Phi(z_1)$ by employing any one of numerous tests such as the Jury/Marden table test outlined above.

Application of, for example, the Jury/Marden table test to (4.52) is straightforward, from a CAD standpoint, given the coefficients. Clearly, however, obtaining these coefficients is not a particularly feasible proposition from the same standpoint. Further, this test does not provide easily used measures of relative stability and/or performance indicators. Hence, as with the tests based on $\rho_D(z)$ written in the form of (4.2) for (a) of theorem 3.3.7 or 3.3.8, the major remit of this test is clearly in low order synthesis type problems where some, or all, of the elements of Φ are design parameters.

Suppose now that (a) and (b) of theorem 3.3.8 hold and then the particular example under consideration is stable along the pass if, and only if, condition (c) holds. Further, consider the interpass transfer-function matrix $G(z_1)$ of (2.102), i.e.

$$G(z_1) = \begin{bmatrix} 0 & I_m & & 0 \\ 0 & & 0 & I_m \\ G_M(z_1) & G_2(z_1) & G_1(z_1) \end{bmatrix} \qquad (4.53)$$

where

$$G_j(z_1) = C(z_1 I_n - \Phi)^{-1} \Delta_{j-1} + D_j, \quad 1 \le j \le M \qquad (4.54)$$

and consider the case of $|z_1| = 1$. Then it follows immediately that this condition is equivalent to the requirement that the continuous curves, or characteristic loci, generated by the eigenvalues $z_j(z_1)$, $1 \le j \le M$, of (4.53) - (4.54) lie entirely within the unit circle in the complex plane for all $|z_1| = 1$. Consequently the testing of this condition reduces to the evaluation and representation of these so-called repetitive system characteristic loci relative to the unit circle in the complex plane. A task which, with the simple additional operation of superimposing the unit circle onto the resulting plots, can be undertaken using standard CAD software for the derived system $L_D(\Phi, \Delta, C, D_0)$.

To develop an alternative test for (c) of theorem 3.3.8 to that given above, first write $\rho(z, z_1) = 0$ with

$$\rho(z, z_1) := |z I_N - G(z_1)| \qquad (4.55)$$

in the form

$$z^N + c_{N-1}(z_1) z^{N-1} + \ldots + c_1(z_1) z + c_0(z_1) = 0 \qquad (4.56)$$

where the coefficients $c_0(z_1)$, $c_1(z_1)$, ..., $c_{N-1}(z_1)$ are rational functions in z_1. Further, let $a_N(z_1)$ denote the least common denominator of the $c_j(z_1)$, $j = 0, 1, \ldots, N - 1$. Then (4.56) can be written as

$$a_N(z_1) z^N + a_{N-1}(z_1) z^{N-1} + \ldots + a_1(z_1) z + a_0(z_1) = 0 \qquad (4.57)$$

where the coefficients $a_0(z_1)$, $a_1(z_1)$, ..., $a_N(z_1)$ are real polynomials in z_1. In which case it follows immediately that (c) of theorem 3.3.8 holds if, and only if, the, assumed irreducible, polynomial

$$\rho(z) = a_N(z_1) z^N + a_{N-1}(z_1) z^{N-1} + \ldots + a_1(z_1) z + a_0(z_1) \qquad (4.58)$$

satisfies

$$\rho(z) \neq 0, \quad \forall z_1: |z_1| = 1, \ |z| \ge 1 \qquad (4.59)$$

To develop a test for (4.59), follow the analysis associated with (4.13) in the differential case and suppose that the N×N Schur-Cohn matrix $H = \{h_{ij}\}$ of (4.14) is constructed from its coefficients. Then in this case each element of H is a polynomial in z_1 and/or its complex conjugate $\bar{z}_1$. Further, it can be shown that (4.59) holds if, and only if, the Hermitian polynomial matrix $H(z_1) \equiv H$ is positive definite $\forall z_1: |z_1| = 1$. Equivalently, for all constant complex vectors $U \neq 0$ of a unitary N-dimensional vector space,

$$U^* H(z_1) U > 0, \quad \forall z_1: |z_1| = 1 \tag{4.60}$$

where * again denotes the complex conjugate transpose.

A Hermitian polynomial matrix $H(z_1)$ satisfying (4.60) is said to be circle positive where this term can be regarded as the dual of the concept of axis positivity introduced in section 4.2. Further, the following result, which is proved using the same arguments as those used in establishing theorem 4.2.1, gives necessary and sufficient conditions for the existence of this property.

<u>Theorem 4.3.1</u>: The Hermitian polynomial matrix $H(z_1)$ is circle positive if, and only if,

    (a) $H(1) > 0$ (4.61)

and

    (b) $|H(z_1)| > 0, \quad \forall z_1: |z_1| = 1$ (4.62)

    The first condition of theorem 4.3.1 requires that the real symmetric matrix $H(1)$ is positive definite. Hence, for example, it holds if, and only if, all principal minors of $H(1)$ are positive. Alternatively, any one of numerous equivalent tests could be employed.

As a first step in developing a test for (4.62), note that $\bar{z}_1 = z_1^{-1}$ on the unit circle and that the determinant of an Hermitian matrix is real. Hence $f(z_1) := |H(z_1)|$ must have the form

$$f(z_1) = \sum_{j=0}^{q} c_j (z_1^j + z_1^{-j}) \tag{4.63}$$

where the coefficients $c_0, c_1, \ldots, c_q$ are real scalars. Consequently (4.62) holds if, and only if,

$$f(z_1) > 0, \quad \forall |z_1| = 1 \tag{4.64}$$

and the following result now expresses this new condition in terms of the roots of $f(z_1)$.

<u>Lemma 4.3.2</u>: The polynomial $f(z_1)$ satisfies (4.64) if, and only if, it has no roots on the unit circle and $f(z_1) > 0$ for some $|z_1| = 1$.

Proof: The absence of roots on the unit circle ensures that $f(z_1)$ is nonzero on this contour and $f(z_1) > 0$ for some $|z_1| = 1$ ensures that it also positive. Note also that this lemma reduces to $f(z_1) \equiv c_0 > 0$ in the trivial case of $q = 0$ and no further analysis is required. ■

The location of the roots of $f(z_1)$ with respect to the unit circle can now be determined as follows. First construct the polynomial $W(z_1)$ as

$$W(z_1) = z_1^q \, f(z_1) = \sum_{j=0}^{q} c_j(z_1^{q+j} + z_1^{q-j}) \qquad (4.65)$$

and note that $W(z_1)$ has the same number of roots on the unit circle as $f(z_1)$ and of the same multiplicities. Hence the following serves as an alternative to lemma 4.3.2.

Lemma 4.3.3: The polynomial $f(z_1)$ satisfies (4.64) if, and only if, $W(z_1)$ has no roots on the unit circle and $W(z_1) > 0$ for some $|z_1| = 1$. ■

Note also that $W(z_1) > 0$ for some $|z_1| = 1$ can be replaced by either $W(1) > 0$ or $W(-1) > 0$, i.e.

$$\sum_{j=0}^{q} c_j > 0 \qquad (4.66)$$

or

$$(-1)^q \sum_{j=0}^{q} (-1)^j c_j > 0 \qquad (4.67)$$

In particular, (4.66) and (4.67) are simple necessary conditions which should be tested before proceeding further.

A polynomial of the form $W(z_1)$ is one example of a so-called self-inversive, or reciprocal, polynomial where this term denotes the fact that the reciprocals of roots inside the unit circle are also roots. Suppose also that $V(W)$ denotes the number of roots of $W(z_1)$ inside the unit circle. Then by the self-inversive property $W(z_1)$ also has $V(W)$ roots outside the unit circle and, since its degree is $2q$,

$$p(W) = 2(q - V(W)) \qquad (4.68)$$

roots on the unit circle. Hence no roots of $W(z_1)$ on the unit circle requires that

$$V(W) = q \qquad (4.69)$$

Further, $V(W)$ can be computed by constructing the Jury/Marden table of (4.47) for $W(z_1)$ and counting the number of positive products of the form (4.46). Note also that since $W(z_1)$ is self-inversive, the polynomial corresponding to $g_1(z)$ in the sequence (4.43) will be identically zero. Consequently in this particular application the first two rows in the table (4.47) must be constructed as per the modification noted previously for the special cases listed under (a) and (b) immediately after (4.49).

To illustrate the use of circle positivity, as expressed by theorem 4.3.1, in the testing of (c) of theorem 3.3.8 consider the case when (after clearing fractions)

$$\rho(z) = (2 z_1^2 + 10 z_1 + 12)z + (z_1^2 + 5 z_1 + 6) \tag{4.70}$$

Then

$$H(z_1) = 18 z_1^2 + 105 z_1 + 186 + 105 z_1^{-1} + 18 z_1^{-2} \tag{4.71}$$

for $|z_1| = 1$ and (4.61) holds since $H(1) > 0$. Further,

$$W(z_1) = 18 z_1^4 + 105 z_1^3 + 186 z_1^2 + 105 z_1 + 18 \tag{4.72}$$

and (4.66) - (4.67) hold. The Jury/Marden table for (4.72) is shown below where the necessary modification has been used to construct the first two rows.

ROW	z_1^0	z_1^1	z_1^2	z_1^3	
1	72	315	373	105	
2	105	373	315	72	
3	$\delta_1 = -5841$	-16380	-6291		
4	-6291	-16380	-5841		
5	$\delta_2 = -5.46 \times 10^6$	-7.37×10^6			
6	-7.37×10^6	-5.46×10^6			
7	$\delta_3 = 24.5 \times 10^6$				(4.73)

Here δ_1 and δ_2 are negative and δ_3 is positive. Hence $V(W) = 2$ and therefore (c) of theorem 3.3.8 holds in this particular case.

At this stage, two alternative tests have been developed for each of the three conditions of theorem 3.3.8. Further, as per theorem 3.3.7 in the differential case of section 4.2, these should be tested in the order of (a) followed by (b) followed by (c) with termination if the one just tested does not hold. Suppose also that the first test is used in each case. Then the following steps represent an eigenvalue, or graphically, based systematic procedure for testing theorem 3.3.8.

STEP 1: Test the necessary condition of (a) by computing the eigenvalues of the matrix D of (4.1) and displaying them relative to the unit circle in the complex plane. Stop if this condition does not hold.

STEP 2: Test (b) by computing the eigenvalues of the matrix Φ and displaying them relative to the unit circle in the complex plane. Stop if this necessary condition does not hold.

STEP 3: Compute the repetitive system characteristic loci generated by the eigenvalues of the interpass transfer-function matrix $G(z_1)$ of (4.53) for all $|z_1| = 1$ and display them relative to the unit circle in the complex plane. The stability along the pass characteristics of the particular example under consideration now follow immediately on visual inspection of the resulting plots.

Using the above systematic procedure, therefore, (a) - (c) of theorem 3.3.8 can be tested using tests expressed in terms of the 2D transfer-function matrix $G(z_1,z)$. (Compare with the first procedure developed for theorem 3.3.7). These tests are suitable for inclusion in a CAD package and are, in effect, well known tests from discrete conventional linear systems theory. To illustrate its use, consider the unit memory process where $Y_{k+1}(P) = X_{k+1}(P)$, $0 \leq P \leq \alpha$, $k \geq 0$, and, under suitable choice of current pass state variables, the dynamics are described by the single equation

$$X_{k+1}(P + 1) = (I_m - A_0^{-1}A_1)X_{k+1}(P) + A_0^{-1}U_{k+1}(P) + \gamma I_m Y_k(P)$$

$$0 \leq P \leq \alpha, \quad X_{k+1}(0) = 0, \quad k \geq 0 \qquad (4.74)$$

where A_0 and A_1 are real constant m×m matrices and γ is a positive real scalar. Further, let $A_0^{-1}A_1$ have eigenvector matrix T and eigenvalues $1 - \eta_1$, $1 - \eta_2, \ldots,$ $1 - \eta_m$ satisfying

$$T^{-1}A_0^{-1}A_1 T = \text{diag}\{1 - \eta_j\}_{1 \leq j \leq m} \qquad (4.75)$$

STEP 1: This step is redundant here since the matrix $D_1 = 0$.

STEP 2: $\Phi = I_m - A_0^{-1}A_1$ in this case and has eigenvalues η_j, $1 \leq j \leq m$. Hence (b) of theorem 3.3.8 holds if, and only if,

$$\max_{1 \leq j \leq m} |\eta_j| < 1 \qquad (4.76)$$

STEP 3: A simple calculation yields that the eigenvalues of $G(z_1) \equiv G_1(z_1)$ are given by

$$z_j(z_1) = \frac{\gamma}{z_1 - \eta_j} \quad , \quad 1 \leq j \leq m \qquad (4.77)$$

Hence, under (4.76), the process of (4.74) is stable along the pass if, and only if, the repetitive system characteristic loci generated by (4.77) lie entirely within the unit circle in the complex plane for all $|z_1| = 1$. Equivalently, the unit circle with centre η_j must lie entirely outside the circle of radius γ and centre the origin, $1 \leq j \leq m$.

Suppose now that each condition in theorem 3.3.8 is tested using the second of the tests developed for it earlier in this section. Then the following steps represent a root clustering, or algebraically based, systematic procedure for

testing this result. This serves as an alternative to the eigenvalue based
systematic procedure detailed above.

STEP 1: Test the necessary condition of (a) by applying any one of numerous
standard tests from discrete conventional linear systems theory to $\rho_D(z)$ of (4.2).
Stop of this condition does not hold.

STEP 2: Test (b) by applying any one of numerous standard tests from discrete
conventional linear systems theory to $\rho_{\Phi}(z_1)$ of (4.51). Stop if this necessary
condition does not hold.

STEP 3: Construct $\rho(z)$ of (4.58) and hence the Schur-Cohn matrix $H(z_1)$ of (4.14).
Test if $H(1)$ is positive definite and stop if this is not the case.

STEP 4: Construct $f(z_1) = |H(z_1)|$ and hence $W(z_1) = z_1^q f(z_1)$ of (4.65). Stop if
(4.66) and (4.67) do not hold for this polynomial.

STEP 5: Test (4.69) by constructing the Jury/Marden table for $W(z_1)$.

As noted when each of them was considered separately, the tests employed in the
above procedure are not suitable for CAD implementation. Hence the major remit of
this procedure is clearly in low order synthesis problems where some, or all, of the
elements of the matrices of the example under consideration are design parameters.

The first of the two systematic test procedures developed in this section uses,
in effect, 'Nyquist like' tests from the stability analysis of $L_D(\Phi,\Delta,C,D_o)$. Hence
this procedure mirrors its differential counterpart of the previous section in terms
of the production of easily used relative stability and/or performance indicators.
This problem is discussed again in section 4.4 and in the next chapter where
alternative simulation-based tests for stability along the pass are developed from
suitably well behaved plant step response data which is assumed to be available.
Further, use of these, sufficient but not necessary, tests produces, at no extra
cost, computable information concerning the rate of approach to the limit profile in
one special case of major practical interest. Finally, chapter 6 considers the use
of this information in the formulation of controller design algorithms.

4.4 Application of 2D Systems Stability Tests

The central result of section 3.4, theorem 3.4.4, states that BIBO stability of
2D linear systems described by the Roesser model of (2.58) is equivalent to
stability along the pass for the discrete unit memory process of example 2.3.8. In
this section, the use of this result in the development of stability tests, and
related topics, for such repetitive processes is considered.

Return, therefore, to the model of example 2.3.8 and regard it as a 2D linear
system described by the Roesser model with 2D transfer-function matrix of the form
(3.137) which is assumed to have no nonessential singularities of the second kind.
Further, see also (3.143), define the following two variable polynomial in terms of
this 2D transfer-function matrix

$$\rho(z_1,z_2) = \begin{vmatrix} I_n - z_1\Phi & -z_1\Delta_0 \\ -z_2 C & I_m - z_2 D_1 \end{vmatrix} \qquad (4.78)$$

Then corollary 3.4.4 is the basic stability result in this context and its testing in all but very simple cases is clearly not a feasible proposition. This fact has led to the development of a large number of equivalent versions of this result, of which corollaries 3.4.5 and 3.4.6 are examples, with varying claims of increased computational efficiency. Generally, all of the resulting tests have been applied to low order SISO problems and the vast majority are root clustering, or algebraically, based. Hence they can be regarded as alternative bases for the unit memory version of the second systematic test procedure developed in section 4.3.

At this stage, define the so-called augmented plant matrix for example 2.3.8 as

$$A = \begin{bmatrix} \Phi & \Delta_0 \\ C & D_1 \end{bmatrix} \qquad (4.79)$$

Then the following result gives three necessary conditions for stability along the pass expressed as stability in the conventional sense of D_1, Φ and A, i.e. all eigenvalues lie in the open unit circle in the complex plane.

Lemma 4.4.1: Consider the extended linear repetitive process $S(E_\alpha, W_\alpha, L_\alpha)_{\alpha \geq \alpha_0}$ generated by the model of example 2.3.8 with $\alpha \geq \alpha_0$. Then the following are necessary conditions for stability along the pass.

(i)   The matrix D_1 is stable.

(ii)   The matrix Φ is stable.

(iii) The matrix A is stable.

Proof: Condition (ii) follows from direct application of (3.145) and (i) follows by reversing the roles of z_1 and z_2 in this result. To establish (iii), consider the particular case of $z_1 = z_2 = 1$ in (3.144). ∎

All of these conditions are easily tested using standard tests from the stability theory of $L_D(\Phi,\Delta,C,D_0)$ and should be established before proceeding further. Note also that (i) is asymptotic stability.

In the development of a systems theory for the derived system, $L_D(\Phi,\Delta,C,D_0)$, the concept of a Lyapunov function has played a very important role. This is one method of stability analysis for $L_D(\Phi,\Delta,C,D_0)$ in the state-space domain and is based on the resulting matrix Lyapunov equation. The maturity and widespread use of this approach in stability analysis and controller design for $L_D(\Phi,\Delta,C,D_0)$ has naturally resulted in work on extending it to similar problems for 2D linear systems described by the Roesser model and hence example 2.3.8.

A study of the published work shows that this general problem has been approached in two essentially different ways. The first of these consists of developing a 2D Lyapunov equation with constant coefficients and the second is based on a 1D Lyapunov equation with coefficients which are functions of a complex variable. In either case, the basic objective is a suitable extension of the Lyapunov theory for $L_D(\Phi,\Delta,C,D_0)$ which gives necessary and sufficient conditions for 2D/repetitive system stability.

The concept of an nD Lyapunov equation was first introduced in the development of a stability theory for an nD continuous, or differential, system, where the problem under study was the development of conditions under which the resulting characteristic polynomial (a function of n variables) is strictly Hurwitz and hence has no zeros (roots) in the region $\text{Re}\{s_i\} \geq 0$, $i = 1,\ldots,n$. This was then extended to the Roesser model using the double bilinear transform to yield a 2D Lyapunov condition for stability. In particular, consider again the augmented plant matrix A of (4.79). Then this condition states that there exists positive definite symmetric matrices Q,W_1,W_2 of dimensions $(n + m) \times (n + m)$, $(n \times n)$ and $(m \times m)$ respectively and $W = W_1 \oplus W_2$ such that the 2D/repetitive process Lyapunov equation

$$A^T W A - W = -Q \tag{4.80}$$

holds where $\oplus$ denotes the direct sum of W_1 and W_2, i.e.

$$W = \begin{bmatrix} W_1 & 0 \\ 0 & W_2 \end{bmatrix} \tag{4.81}$$

Equation (4.80) has constant coefficients but, unlike its conventional linear systems counterpart, is, in general, only a sufficient condition for 2D/repetitive system stability. This is a well established result (see, for example, the cited reference) and hence the proof is omitted here except to note that it is based on the concept of a strictly bounded real matrix from circuit theory. There are, however, a number of special cases when it is necessary and sufficient and the following analysis considers the two of these which are most in keeping with the general aim of this chapter.

Suppose that the example under consideration is SISO. Then the following lemma, which follows immediately as a special case of the general result and is hence stated without proof, shows that (4.80) is necessary and sufficient in this case.

Lemma 4.4.2: Consider the SISO version of the model of example 2.3.8 and suppose that the pairs (Φ,C) and (Φ,Δ_0^T) are completely reachable and observable respectively. Then the extended linear repetitive process $S(E_\alpha,W_\alpha,L_\alpha)_{\alpha \geq \alpha_0}$ generated by this model with $\alpha \geq \alpha_0$ is stable along the pass if, and only if, there exists symmetric positive definite matrices $W_1, W_2, W = W_1 \oplus W_2$ and Q such that (4.80) holds. ■

The second special case of interest here is when A of (4.79) is normal and hence

$$A^T A = A A^T \qquad (4.82)$$

Further, note that (4.80) is equivalent to

$$A^T W A - W < 0 \qquad (4.83)$$

where the inequality denotes negative definiteness. This is structurally similar to the Lyapunov equation for $L_D(\Phi, \Delta, C, D_0)$. Hence it follows that stability along the pass under (4.83) implies that A is a stability matrix in the conventional sense. This is now strengthened to a necessary and sufficient condition for stability along the pass by the following result. Equivalently, (iii) of lemma 4.4.1 is necessary and sufficient in this special case.

Theorem 4.4.1: Suppose that the matrix A of (4.79) for the discrete unit memory linear repetitive process of example 2.3.8 is normal. Then the extended linear repetitive process $S(E_\alpha, W_\alpha, L_\alpha)_{\alpha \geq \alpha_0}$ generated by the model of this example with $\alpha \geq \alpha_0$ is stable along the pass if, and only if, there exists symmetric positive definite matrices W_1, W_2, $W = W_1 \oplus W_2$ such that (4.83) holds.

Proof: Necessity is immediate from the above discussion. To prove sufficiency, first note that if A is stable in the conventional sense then it satisfies the conventional linear systems Lyapunov equation, i.e.

$$A^T W A - W < 0 \qquad (4.84)$$

Further, denote the eigenvalues of A by λ_i, $1 \leq i \leq n + m$, and the corresponding eigenvector matrix by R. Then, since A is normal,

$$A = R \Lambda R^* \qquad (4.85)$$

where * again denotes the complex conjugate transpose and $\Lambda = \text{diag}\{\lambda_i\}_{1 \leq i \leq n+m}$

Substituting (4.85) and $W = I_{n+m}$ into the left hand side of (4.84) now yields

$$A^T W A - W = R(|\Lambda|^2 - I_{n+m})R^* \qquad (4.86)$$

where $|\Lambda|^2 = \text{diag}\{|\lambda_i|^2\}_{1 \leq i \leq n+m}$, and this matrix is negative definite since $|\lambda_i|^2 < 1$, $1 \leq i \leq n + m$, by the assumption that A is stable in the conventional sense. Equivalently, (4.84) (the conventional linear systems case) holds under the choice of $W = I_{n+m}$ and, since I_{n+m} is block diagonal under any partition, A also satisfies the 2D/repetitive process Lyapunov equation (4.83) and hence stability along the pass. ■

Suppose, therefore, that A of (4.79) is normal for the particular example under consideration. Then, in effect, theorem 4.4.1 states that stability along the pass can be determined by testing A for stability in the conventional sense. Further, this result can be extended to a number of other sub-classes by use of appropriate similarity transformations. The details of these can be found in the cited reference.

Return now to the general case where (4.80) is a sufficient, but not necessary, condition for stability along the pass. Then this fact clearly reduces its effectiveness in this context given the necessary and sufficient tests of section 4.3. In either of the special cases detailed here, however, it could be used as an alternative route to testing for stability along the pass. This is particularly relevant for the test of theorem 4.4.1 which is clearly computationally more efficient than either of the systematic test procedures developed in section 4.3. Further, note that numerous well defined algorithms exist for computing positive definite solutions of this 2D Lyapunov equation. Such algorithms have arisen due to their importance in other problem areas such as the design of low-noise and/or limit cycle free 2D filter realisations, the computation of grammians for 2D model reduction and (possibly) stability testing for nonlinear 2D systems. (See the cited references for complete details.)

One use of this constant coefficient Lyapunov equation in the 2D systems context has been in the development of stability margins with a view to use in control systems design. This is an active research area and numerous publications have appeared on various aspects. The extension of this concept to example 2.3.8 is examined later in this section after the second version of the Lyapunov equation has been considered.

The second version of a Lyapunov equation for a 2D/repetitive process is, in effect, a 1D, or conventional linear systems, Lyapunov equation with coefficients which are functions of a complex variable. It is, in effect, based on the following set of necessary and sufficient conditions for stability along the pass (given (i) of lemma 4.4.1 by assumption):

(a)   the matrix Φ is stable in the conventional sense; and

(b)   the transfer-function matrix
$$G_1(z_1^{-1}) = C(z_1^{-1}I_n - \Phi)^{-1}\Delta_0 + D_1 \tag{4.87}$$
with $|z_1^{-1}| = 1$ is stable in the conventional sense.

Consider now the unit circle in the z_1^{-1} plane $C_1 = \{z_1^{-1}: |z_1^{-1}| = 1\}$. Then for each fixed $z_1^{-1} \in C_1$, $G_1(z_1^{-1})$ is, in general, a complex matrix. Further, the following result gives a condition for an arbitrary matrix with a complex parameter, say $F(z)$, to be stable for each $|z| = 1$ in the conventional sense.

Theorem 4.4.2: Consider a matrix $F(z)$ where the complex parameter z satisfies $|z| = 1$. Then this matrix is stable in the conventional sense if, and only if, for any given positive definite Hermitian, denoted P.D.H., matrix $W(z)$ with $|z| = 1$ there exists a unique P.D.H. matrix $H(z)$ such that
$$H - F^*H F = W \tag{4.88}$$

where * again denotes the complex conjugate transpose.

<u>Proof:</u> Consider first sufficiency and take a fixed $|z| = 1$. Further, let λ and ν be any eigenvalue and eigenvector of F respectively, i.e.

$$F\nu = \lambda\nu \qquad (4.89)$$

and

$$\nu^* F^* = \bar{\lambda}\nu^* \qquad (4.90)$$

Pre-multiplying (4.88) by ν^* and post-multiplying it by ν now yields

$$\nu^* H\nu - \nu^* F^* HF\nu = \nu^* H\nu(1 - |\lambda|^2) = \nu^* W\nu \qquad (4.91)$$

Hence

$$1 - |\lambda|^2 = \frac{\nu^* W\nu}{\nu^* H\nu} > 0 \qquad (4.92)$$

and therefore F is stable.

To prove necessity, suppose that F is stable for any fixed $|z| = 1$ and let a P.D.H. matrix W be given. In which case consider

$$H = \sum_{q=0}^{\infty} (F^*)^q W F^q \qquad (4.93)$$

and note that this H is well defined since F is stable and is clearly a P.D.H. matrix. Further,

$$H - F^* H F = \sum_{q=0}^{\infty} (F^*)^q W F^q - \sum_{q=0}^{\infty} (F^*)^{q+1} W F^{q+1} = W \qquad (4.94)$$

and suppose that H_1 is another solution of (4.88). Then

$$H = \sum_{q=0}^{\infty} (F^*)^q W F^q = \sum_{q=0}^{\infty} (F^*)^q (H_1 - F^* H_1 F) F^q$$

$$= \sum_{q=0}^{\infty} (F^*)^q H_1 F^q - \sum_{q=0}^{\infty} (F^*)^{q+1} H_1 F^{q+1} = H_1 \qquad (4.95)$$
∎

Using this result now gives the following set of conditions for stability along the pass of example 2.3.8 which are equivalent to corollary 3.3.11.

<u>Theorem 4.4.3:</u> The extended linear repetitive process $S(E_\alpha, W_\alpha, L_\alpha)_{\alpha \geq \alpha_0}$ generated by the model of example 2.3.8 with $\alpha \geq \alpha_0$ is stable along the pass if, and only if,

(a) the matrix D_1 is stable in the conventional sense;

(b) the matrix Φ is stable in the conventional sense; and

(c) the matrix equation

$$H(z_1^{-1}) - G_1^*(z_1^{-1})H(z_1^{-1})G_1(z_1^{-1}) = W(z_1^{-1}) \qquad (4.96)$$

has a P.D.H. solution $H(z_1^{-1})$ for any given P.D.H. matrix $W(z_1^{-1})$ and any $z_1^{-1} \in C_1$.
∎

Recall now the unit memory versions of the systematic test procedures developed in section 4.3. Then, in effect, theorem 4.4.3 serves as an alternative to either of these procedures. Further, these new conditions should obviously be tested in the order of (a) followed by (b) followed by (c) with termination if the one just tested does not hold. An additional step before proceeding to (c) being to examine the necessary condition listed under (c) in lemma 4.4.1, i.e. stability in the conventional sense of the augmented plant matrix A.

The testing of (a) and (b) of theorem 4.4.3 is just the standard stability problem for $L_D(\Phi, \Delta, C, D_0)$. To develop a test for (c), suppose that $W(z_1^{-1})$ is given where $W(z_1^{-1}) = I_m$ is an obvious choice. In which case (4.96) can be solved to yield a rational matrix solution $H(z_1^{-1})$ which is obviously Hermitian. Further, write z_1^{-1} in polar form as $z_1^{-1} = e^{i\theta}$, $0 \le \theta \le 2\pi$, and denote the j-th order principal minor of $H(z_1^{-1})$ by $h_j(\theta)$, $1 \le j \le m$. Then the $h_j(\theta)$ are functions of the single real variable θ over the closed interval $[0, 2\pi]$ and the following corollary is effectively a systematic test procedure for theorem 4.4.3.

Corollary 4.4.3: The conditions of theorem 4.4.3 are equivalent to the following:

    (a)   the matrix D_1 is stable in the conventional sense;

    (b)   the matrix Φ is stable in the conventional sense; and

    (c) $h_j(\theta) > 0$, $0 \le \theta \le 2\pi$, $1 \le j \le m$ (4.97)

To illustrate the direct use of this corollary in a synthesis problem, consider the SISO case where D_1 (a scalar in this case) satisfies $|D_1| < 1$ and hence asymptotic stability. Suppose also that Φ has eigenvalue-eigenvector decomposition

$$T_2^{-1} \Phi \, T_2 = \text{diag}\{\lambda_j\}_{1 \le j \le n} \qquad (4.98)$$

Then (b) holds if, and only if, $|\lambda_j| < 1$, $1 \le j \le n$.

Suppose now that (b) holds and consider the case when λ_j, $1 \le j \le n$, is real. Further, define T as $T = T_2 \oplus 1$ and transform the augmented plant matrix to

$$\tilde{A} \equiv T^{-1}AT = \begin{bmatrix} \Lambda & \tilde{A}_2 \\ \tilde{A}_3 & D_1 \end{bmatrix} \qquad (4.99)$$

where

$$\tilde{A}_2 = T_2^{-1} \Delta_0 \equiv (d_1, \ldots, d_n)^T \qquad (4.100)$$

and

$$\tilde{A}_3 = CT_2 \equiv (f_1, \ldots, f_n) \qquad (4.101)$$

Now define

$$\delta_i = d_i f_i, \quad 1 \le i \le n \qquad (4.102)$$

and consider the special case when all of these numbers have the same sign. Then, as shown below, (c) holds if, and only if,

$$\max\{|D_1 + C(I_n - \Phi)^{-1}\Delta_0|, |D_1 - C(I_n + \Phi)^{-1}\Delta_0|\} < 1 \tag{4.103}$$

To prove (4.103), take $W(e^{i\theta}) \equiv 1$ in (4.96) to yield $H(e^{i\theta}) \equiv h_1(\theta)$ with $z_1^{-1} = e^{i\theta}$ from

$$h_1(\theta)(1 - |D_1 + C(z_1^{-1}I_n - \Phi)^{-1}\Delta_0|^2) = 1 \tag{4.104}$$

Hence $h_1(\theta) > 0$, $0 \le \theta \le 2\pi$, if, and only if,

$$|D_1 + C(z_1^{-1}I_n - \Phi)^{-1}\Delta_0| < 1, \quad z_1^{-1} = e^{i\theta} \tag{4.105}$$

and necessity is immediate. For sufficiency, note that

$$
\begin{aligned}
|\eta(z_1^{-1})| &:= |D_1 + C(z_1^{-1}I_n - \Phi)^{-1}\Delta_0| \\
&= |D_1 + \tilde{A}_3(z_1^{-1}I_n - \Lambda)^{-1}\tilde{A}_2| \\
&= |D_1 + \sum_{i=1}^{n} \frac{\delta_i}{z_1^{-1} - \lambda_i}| \\
&= |D_1 + \text{sgn}(\delta_1) \sum_{i=1}^{n} \eta_i(z_1^{-1})| \tag{4.106}
\end{aligned}
$$

where

$$\eta_i(z_1^{-1}) = \frac{|\delta_i|}{z_1^{-1} - \lambda_i}, \qquad 1 \le i \le n \tag{4.107}$$

and each of these functions maps the unit circle in the z_1^{-1} plane onto a circle centred on the real line. Hence the maximum value can only occur when $z_1^{-1} = 1$ or -1 and (4.103) follows immediately.

Note: A similar analysis can be used to provide a sufficient condition for the case when the δ_i of (4.102) have different signs.

Return now to the problem of testing (c) in the general case. Then it is clear that obtaining the $h_j(\theta)$ and testing each of them for positive definiteness could lead to very severe computational expense. The following analysis uses the Kronecker product, denoted $\otimes$, to obtain an equivalent result to theorem 4.4.3 whose conditions are expressed in terms of the eigenvalues of constant matrices. This is a two stage operation as detailed below.

Suppose, therefore, that a P.D.H. matrix $W(e^{i\theta})$ is given. Then it is required to show the existence of a P.D.H. matrix $H(e^{i\theta})$ which solves the following equivalent version of (4.96) $\forall \theta \in [0, 2\pi]$.

$$H(e^{i\theta}) - G_1^T(e^{-i\theta})H(e^{i\theta})G_1(e^{i\theta}) = W(e^{i\theta}) \tag{4.108}$$

Further, use of the Kronecker product enables (4.108) to be rewritten as

$$(I_{m^2} - G_1^T(e^{-i\theta}) \otimes G_1^T(e^{i\theta}))S[H(e^{i\theta})] = S[W(e^{i\theta})] \tag{4.109}$$

where $S[.]$ denotes the stacking operator. This now yields the following equivalent result to theorem 4.4.3 which then leads to theorem 4.4.5 below whose conditions are expressed in terms of the eigenvalues of constant matrices.

<u>Theorem 4.4.4:</u>   The conditions of theorem 4.4.3 are equivalent to the following:

(a)   the matrix D_1 is stable in the conventional sense;

(b)   the matrix Φ is stable in the conventional sense;

(c) $H \equiv H(e^{i\theta_0})$, the solution of

$$H - G_1^T(e^{-i\theta_0})H \, G_1(e^{i\theta_0}) = W \qquad (4.110)$$

is positive definite for any given positive definite matrix W and an arbitrary $\theta_0 \in [0, 2\pi]$; and

(d) $|I_{m^2} - G_1^T(e^{-i\theta}) \otimes G_1^T(e^{i\theta})| \neq 0, \quad \forall \, \theta \in [0, 2\pi]$ \qquad (4.111)

<u>Proof:</u>   It is clearly required to show that (c) and (d) above are, together, equivalent to (c) of theorem 4.4.3. First note, therefore, that (4.111) guarantees the existence of a unique solution, $H(e^{i\theta})$, of (4.109). Further, $H(e^{i\theta})$ is P.D.H. if, and only if, its eigenvalues are positive $\forall \, \theta \in [0, 2\pi]$. These are continuous functions of θ and will always be positive if $H(e^{i\theta_0})$ is positive definite for an arbitrary θ_0 and (4.111) holds. Hence (4.110) and (4.111) are equivalent to (c) of theorem 4.4.3 and proof is complete. ∎

Consider now the problem of testing theorem 4.4.4 for a given example. Then, in effect, this consists of testing three constant matrices, D_1, Φ and $H(e^{i\theta_0})$, for stability in the conventional sense and testing (4.111) $\forall \, \theta \in [0, 2\pi]$. Hence it is clear that using this result as a basis for a systematic test procedure will not lead to increased computational efficiency. The following theorem now expresses (4.111) in terms of the eigenvalues of constant matrices.

<u>Theorem 4.4.5:</u>   The conditions of theorem 4.4.4 are equivalent to the following:

(a)   the matrix D_1 is stable in the conventional sense;

(b)   the matrix Φ is stable in the conventional sense;

(c)   the matrix $H(e^{i\theta_0})$ is stable in the conventional sense for an arbitrary θ_0 $\in [0, 2\pi]$; and

(d) $|\lambda^2 X_1 + \lambda X_2 + X_3| \neq 0, \quad \forall \, |\lambda| = 1$ \qquad (4.112)

where

$$X_1 = \begin{bmatrix} 0 & 0 & 0 & 0 \\ 0 & 0 & 0 & 0 \\ 0 & 0 & 0 & 0 \\ 0 & 0 & 0 & -\Phi^T \otimes I_n \end{bmatrix} \qquad (4.113)$$

$$X_2 = \begin{bmatrix} 0 & \Delta_0^T \otimes D_1^T & 0 & \Delta_0^T \otimes \Delta_0^T \\ 0 & -\Phi^T \otimes I_m & 0 & 0 \\ 0 & 0 & I_{mn} & 0 \\ 0 & 0 & 0 & I_{n^2} + \Phi^T \otimes \Phi^T \end{bmatrix}$$

(4.114)

$$X_3 = \begin{bmatrix} I_{m^2} - D_1^T \otimes D_1^T & 0 & I_m \otimes \Delta_0^T & 0 \\ C^T \otimes I_m & I_{mn} & 0 & 0 \\ D_1^T \otimes C^T & 0 & -I_m \otimes \Phi^T & 0 \\ C^T \otimes C^T & 0 & 0 & -I_n \otimes \Phi^T \end{bmatrix}$$

(4.115)

Proof: In the case of (a) - (c), it is clearly only necessary to note that $H(e^{i\theta_0})$ is the solution of the conventional linear systems Lyapunov equation and hence all its eigenvalues have modulus strictly less than unity. Consequently it remains to prove the equivalence of (d) in each case.

Suppose, therefore, that

$$f(e^{i\theta}) = |I_{m^2} - G_1^T(e^{-i\theta}) \otimes G_1^T(e^{i\theta})|$$

(4.116)

Then

$$f(e^{i\theta}) = |I_{m^2} - D_1^T \otimes D_1^T - e^{i\theta}(\Delta_0^T \otimes D_1^T)((I_n - e^{i\theta}\Phi^T) \otimes I_m)^{-1}(C^T \otimes I_m)$$
$$- (I_m \otimes \Delta_0^T)(I_m \otimes (I_n e^{i\theta} - \Phi^T))^{-1}(D_1^T \otimes C^T)$$
$$- e^{i\theta}(\Delta_0^T \otimes \Delta_0^T)g(e^{i\theta})^{-1}(C^T \otimes C^T)|$$

(4.117)

where the matrix $g(e^{i\theta})$ whose inverse appears in the last term is given by

$$g(e^{i\theta}) = -(\Phi^T \otimes I_n)e^{i2\theta} + (I_{n^2} + \Phi^T \otimes \Phi^T)e^{i\theta} - I_n \otimes \Phi^T$$

(4.118)

Hence

$$f(e^{i\theta}) = |I_{m^2} - D_1^T \otimes D_1^T - U^T V^{-1} W|$$

(4.119)

where

$$U^T = [e^{i\theta}(\Delta_0^T \otimes D_1^T), \ (I_m \otimes \Delta_0^T), \ e^{i\theta}(\Delta_0^T \otimes \Delta_0^T)]$$

(4.120)

$$V^{-1} = \begin{bmatrix} I_{mn} - e^{i\theta}\Phi^T \otimes I_m & 0 & 0 \\ 0 & I_{mn}e^{i\theta} - I_m \otimes \Phi^T & 0 \\ 0 & 0 & g(e^{i\theta}) \end{bmatrix}^{-1}$$

(4.121)

and

$$W = \begin{bmatrix} C^T \otimes I_m \\ D_1^T \otimes C^T \\ C^T \otimes C^T \end{bmatrix} \qquad (4.122)$$

Further, (b) implies that

$$|I_{mn} - e^{i\theta}\Phi^T \otimes I_m||I_{mn}e^{i\theta} - I_m \otimes \Phi^T||g(e^{i\theta})| \neq 0, \quad \forall\ \theta \in [0,2\pi] \qquad (4.123)$$

and suppose that $f(e^{i\theta})$ is pre-multiplied by the left hand side of (4.123). In which case it follows immediately that

$$f(e^{i\theta}) \neq 0, \quad \forall\ \theta \in [0,2\pi] \qquad (4.124)$$

is equivalent to

$$|X| \neq 0, \quad \forall\ \theta \in [0,2\pi] \qquad (4.125)$$

where

$$X = \begin{bmatrix} I_{m^2} - D_1^T \otimes D_1^T & e^{i\theta}(\Delta_0^T \otimes D_1^T) & I_m \otimes \Delta_0^T & e^{i\theta}(\Delta_0^T \otimes \Delta_0^T) \\ C^T \otimes I_m & I_{mn} - e^{i\theta}\Phi^T \otimes I_m & 0 & 0 \\ D_1^T \otimes C^T & 0 & I_{mn}e^{i\theta} - I_m \otimes \Phi^T & 0 \\ C^T \otimes C^T & 0 & 0 & g(e^{i\theta}) \end{bmatrix} \qquad (4.126)$$

Finally, it is clear that (4.126) and (4.112) are identical and the proof is complete. ∎

To examine theorem 4.4.5 for a given example, it is necessary to test three constant matrices for stability in the conventional sense and the second order matrix polynomial condition of (4.112). Note also that the matrix X_1 of (4.113) is singular and hence further development is required for the two separate cases when Φ is singular and non-singular. In particular, extensive, but routine, algebraic manipulations must be performed in both cases to reformulate (4.112) as a condition involving a first order matrix polynomial which can be easily tested with existing software for computing generalised eigenvalues. The details can be found in the cited reference.

Consider now the problem of testing example 2.3.8 for stability along the pass. Then it is clear that the obvious systematic procedure for testing theorem 4.4.5 serves as an alternative to either of the systematic procedures of section 4.3. Detailed comparative studies would, however, require the results from application of all of these procedures to suitably defined benchmark problems and this topic is not considered further here.

To date, no consideration has been given in this work to the development of any form, CAD orientated or otherwise, of stability margins. The following analysis represents a first attempt at this wide ranging problem for processes described by example 2.3.8. In particular, it extends some work from 2D linear systems described by the Roesser model to this case and briefly discusses the potential for further developments in this area.

Return, therefore, to the case of a constant coefficient Lyapunov equation for example 2.3.8. In particular, suppose that there exists suitably dimensioned symmetric positive definite matrices Q, W_1, W_2 and $W = W_1 \oplus W_2$ such that the 2D/repetitive process Lyapunov equation of (4.80), i.e.

$$W - A^T W A = Q \tag{4.127}$$

holds where the augmented plant matrix A is defined by (4.79). Then this is, in general, a sufficient condition for stability along the pass and hence $\rho(z_1, z_2)$ of (4.78) satisfies

$$\rho(z_1, z_2) \neq 0 , \quad \forall |z_1| \leq 1, \ |z_2| \leq 1 \tag{4.128}$$

Suppose now that (4.128) holds and consider the 2D linear systems case. Then here the stability margin has been introduced as a criterion for characterising the spatial domain performance of such systems. It is defined using the largest bidisc where $\rho(z_1, z_2)$ has no roots, i.e.

$$\rho(z_1, z_2) \neq 0 \text{ in } U_{\sigma1}^2 = \{(z_1, z_2): |z_1| < 1 + \sigma_1, \ |z_2| < 1\} \tag{4.129}$$

$$\rho(z_1, z_2) \neq 0 \text{ in } U_{\sigma2}^2 = \{(z_1, z_2): |z_1| < 1, \ |z_2| < 1 + \sigma_2\} \tag{4.130}$$

$$\rho(z_1, z_2) \neq 0 \text{ in } U_\sigma^2 = \{(z_1, z_2): |z_1| < 1 + \sigma, \ |z_2| < 1 + \sigma\} \tag{4.131}$$

Considerable effort has also been directed towards the development of algorithms for computing σ_1, σ_2 and σ. This has yielded numerous algorithms based on different approaches. For example, one set is based on minimising the distance between the roots of $\rho(z_1, z_2)$ and the boundary of the unit bidisc $T^2 = \{(z_1, z_2): |z_1| = 1, |z_2| = 1\}$. Alternatively algorithms based on the so-called resultant matrix could be used. Further, it has been shown that the stability margin is related (in a well defined sense) to the minimal norm of the augmented plant matrix.

In the 2D case, it is not always necessary to know the exact value of the stability margin. Instead, it suffices to know that it is greater than certain lower limits where one such limit, or bound, can be obtained as a function of the positive definite solution to the Lyapunov equation (4.127). The analysis which follows extends this method to example 2.3.8 and discusses the outcome in terms of systems analysis.

Consider, therefore, the case when suitably dimensioned symmetric positive definite solutions W and Q exist for the Lyapunov equation (4.127) where $W = W_1 \oplus W_2$ and

$$Q = \begin{bmatrix} Q_1 & Q_2 \\ Q_2^T & Q_3 \end{bmatrix} \tag{4.132}$$

Then this is, in general, a sufficient condition for stability along the pass and hence (4.128) holds. Further, the following analysis yields a lower bound for the stability margin as a function of the matrices W_1, W_2, Q_1, Q_2 and Q_3. First pre and post-multiply (4.127) by the matrix $\beta_1 I_n \oplus \beta_2 I_m$ where β_1 and β_2 are positive real scalars and add W to both sides of the result. This yields, after some algebraic manipulations,

$$\begin{bmatrix} W_1 & 0 \\ 0 & W_2 \end{bmatrix} - \begin{bmatrix} \beta_1 \Phi & \beta_2 \Delta_o \\ \beta_1 C & \beta_2 D_1 \end{bmatrix}^T \begin{bmatrix} W_1 & 0 \\ 0 & W_2 \end{bmatrix} \begin{bmatrix} \beta_1 \Phi & \beta_2 \Delta_o \\ \beta_1 C & \beta_2 D_1 \end{bmatrix} = \hat{Q} \tag{4.133}$$

where

$$\hat{Q} = \begin{bmatrix} \beta_1^2 Q_1 + (1 - \beta_1^2) W_1 & \beta_1 \beta_2 Q_2 \\ \beta_1 \beta_2 Q_2^T & \beta_2^2 Q_3 + (1 - \beta_2^2) W_2 \end{bmatrix} \tag{4.134}$$

Suppose also that $\hat{Q}$ is positive definite and hence, since W is assumed positive definite, a sufficient condition holds for

$$\hat{\rho}(z_1, z_2) \neq 0 , \ \forall \ |z_1| \leq 1, \quad |z_2| \leq 1 \tag{4.135}$$

where

$$\hat{\rho}(z_1, z_2) = \begin{vmatrix} I_n - z_1 \beta_1 \Phi & -z_1 \beta_2 \Delta_o \\ -z_2 \beta_1 C & I_m - z_2 \beta_2 D_1 \end{vmatrix} \tag{4.136}$$

The following result now establishes the relationship between the roots of $\rho(z_1, z_2)$ and $\hat{\rho}(z_1, z_2)$.

<u>Lemma 4.4.1:</u> Let (z_1, z_2) and $(\hat{z}_1, \hat{z}_2)$ denote the roots of $\rho(z_1, z_2)$ and $\hat{\rho}(z_1, z_2)$ respectively. Then

$$(\hat{z}_1, \hat{z}_2) = (\beta_1^{-1} z_1, \beta_2^{-1} z_2) \tag{4.137}$$

Proof: By definition

$$\hat{\rho}(z_1, z_2) = \begin{vmatrix} I_n - z_1 \beta_1 \Phi & -z_1 \beta_2 \Delta_o \\ -z_2 \beta_1 C & I_m - z_2 \beta_2 D_1 \end{vmatrix}$$

$$= \beta_1^n \beta_2^m \begin{vmatrix} \beta_1^{-1} I_n - z_1 \Phi & -z_1 \Delta_o \\ -z_2 C & \beta_2^{-1} I_m - z_2 D_1 \end{vmatrix}$$

$$= \begin{vmatrix} I_n - z_1\beta_1\Phi & -z_1\beta_1\Delta_0 \\ \\ -z_2\beta_2 C & I_m - z_2\beta_2 D_1 \end{vmatrix}$$

$$= \rho(\beta_1 z_1, \beta_2 z_2) \tag{4.138}$$

and (4.137) follows immediately. ∎

Using this result, it is possible to characterise the locations of the roots of $\hat{\rho}(z_1, z_2)$ as functions of β_1 and β_2. In particular, if $\beta_1 = \beta_2 = 1$ then it is obvious that $\rho(z_1, z_2)$ and $\hat{\rho}(z_1, z_2)$ are identical and satisfy (4.128). If, however, $\beta_i < 1$, $i = 1,2$, the roots of $\hat{\rho}(z_1, z_2)$ move from (z_1, z_2) towards infinity and for $\beta_i > 1$ they move towards the boundary, T^2, of the unit bidisc and, eventually, some of them will be inside this bidisc. Further, note again that if $\hat{Q}$ of (4.134) is positive definite for a given pair (β_1, β_2) then this is a sufficient condition for the corresponding $\hat{\rho}(z_1, z_2)$ to satisfy (4.135). Consequently the range of (β_1, β_2) for which $\hat{Q}$ remains positive definite is closely related to the distance between the roots of $\rho(z_1, z_2)$ and T^2. Further, given that this is in general a sufficient, but not necessary, condition, it follows immediately that the range of (β_1, β_2) for which $\hat{Q}$ is positive definite can only give lower bounds, not actual values, for the stability margin. The analysis below obtains lower bounds for σ_1, σ_2 and σ of (4.129)-(4.131) respectively in terms of the range of (β_1, β_2) for which $\hat{Q}$ is positive definite.

Considering first σ_1, it is clear that a lower bound in this case can be obtained from the range of β_1 for which $\hat{Q}$ is positive definite with $\beta_2 = 1$. This is equivalent to

$$Q_3 > 0 \tag{4.139}$$

and

$$\beta_1^2 Q_1 + (1 - \beta_1^2)W_1 - \beta_1^2 Q_2 Q_3^{-1} Q_2^T > 0 \tag{4.140}$$

Further, (4.139) holds by assumption and hence it remains to consider (4.140) which is equivalent to

$$(W_1 - Q_1 + Q_2 Q_3^{-1} Q_2^T) - \beta_1^2 W_1 < 0 \tag{4.141}$$

and the values of β_1 for which this new condition holds can be obtained from the established theory of the extremal properties of pencils of quadratic forms with the structure $F - \lambda B$. In particular, it is known that for $B > 0$

$$\lambda_{min}[B^{-1}F] \leq \frac{x^T F x}{x^T B x} \leq \lambda_{max}[B^{-1}F], \quad x \neq 0 \tag{4.142}$$

and

$$F - \lambda_{max}[B^{-1}F]B \leq 0 \qquad (4.143)$$

where $\lambda_{min}[B^{-1}F]$ and $\lambda_{max}[B^{-1}F]$ denote the minimum and maximum eigenvalues of $B^{-1}F$ respectively (and $\leq$ in (4.143) denotes the fact that the matrix on the left-hand side is negative semi-definite). Suppose also that B and F are defined by

$$B = W_1 \qquad (4.144)$$

and

$$F = W_1 - Q_1 + Q_2Q_3^{-1}Q_2^T \qquad (4.145)$$

where $W_1 > 0$ by assumption. Then the following lemma shows that this particular choice of F is positive semi-definite (written $F \geq 0$).

Lemma 4.4.2: Suppose that W_1, W_2, Q_1, Q_2 and Q_3 are the (appropriately) dimensioned matrices of a positive definite solution to the Lyapunov equation (4.127). Then F of (4.145) is positive semi-definite

Proof: Rewrite (4.127) as

$$A^TWA = \begin{bmatrix} W_1 - Q_1 & -Q_2 \\ -Q_2^T & W_2 - Q_3 \end{bmatrix} \qquad (4.146)$$

where $W > 0$ and hence the right-hand side is positive semi-definite. This implies that

$$W_1 - Q_1 \geq 0 \qquad (4.147)$$

and $Q_3 > 0$ since $Q > 0$. Hence

$$Q_2Q_3^{-1}Q_2^T \geq 0 \qquad (4.148)$$

and $F \geq 0$ follows immediately from (4.147) and (4.148). ∎

Given this result, (4.142) and (4.143) can now be used to obtain the values of β_1 for which (4.141) holds. This yields

$$\beta_1^{-1} < \sqrt{\lambda_{max}[C_1]} \qquad (4.149)$$

where

$$C_1 = B^{-1}F = (I_n - W_1^{-1}Q_1 + W_1^{-1}Q_2Q_3^{-1}Q_2^T) \qquad (4.150)$$

and all eigenvalues of this matrix are real and non-negative since $F \geq 0$ and $B > 0$. Further, the maximum eigenvalue of C_1 is positive for an augmented plant matrix $A \neq 0$ and the lower bound for σ_1 is now given by

$$\sigma_1 \geq (\sqrt{\lambda_{max}[C_1]})^{-1} - 1 \qquad (4.151)$$

In the case of a lower bound for σ_2 of (4.130), a completely analogous analysis to that above yields

$$\sigma_2 \geq (\sqrt{\lambda_{max}[C_2]})^{-1} - 1 \qquad (4.152)$$

where
$$C_2 = I_m - W_2^{-1}Q_3 + W_2^{-1}Q_2^TQ_3^{-1}Q_2 \tag{4.153}$$

Similarly, it is clear that a lower bound for σ of (4.131) can be obtained by setting $\beta_1 = \beta_2 = \beta$ and determining the values of this parameter for which $\hat{Q}$ of (4.134) is positive definite. This implies that
$$\hat{Q} = W - \beta^2(W-Q) > 0 \tag{4.154}$$

or
$$(W - Q) - \beta^{-2}W < 0 \tag{4.155}$$

The matrix W is positive definite and $(W - Q)$ is positive semi-definite, which follows immediately from writing (4.127) as
$$A^TWA = W - Q \tag{4.156}$$

Hence the range of β for which (4.155) holds can be determined in a similar manner to, for example, β_1 of (4.149). This yields
$$\beta^{-1} < \sqrt{\lambda_{max}[C_3]} \tag{4.157}$$

where
$$C_3 = I_{n+m} - W^{-1}Q \tag{4.158}$$

and hence the lower bound for σ as
$$\sigma \geq (\sqrt{\lambda_{max}[C_3]})^{-1} - 1 \tag{4.159}$$

To illustrate these bounds, consider the special case when
$$A = \begin{bmatrix} -0.5 & -0.395 \\ 1 & -0.01 \end{bmatrix} \tag{4.160}$$

Then the Lyapunov equation (4.127) has the following solution
$$W = \begin{bmatrix} 1 & 0 \\ 0 & 0.395 \end{bmatrix}, \quad Q = \begin{bmatrix} 0.355 & -0.194 \\ -0.194 & 0.239 \end{bmatrix} \tag{4.161}$$

and hence (4.150) and (4.151) yield
$$\sigma_1 \geq 0.116 \tag{4.162}$$

Further, (4.152) and (4.153) yield
$$\sigma_2 \geq 0.127 \tag{4.163}$$

and from (4.154) it follows that
$$W - \beta^2(W - Q) > 0 \tag{4.164}$$

for $\beta < 1.083$ and hence
$$\sigma \geq 0.083 \tag{4.165}$$

As a comparison, the exact values of σ_1, σ_2 and σ, obtained using algorithms detailed in the cited references, are
$$\sigma_1 = 0.12, \quad \sigma_2 = 0.282, \quad \sigma = 0.108 \tag{4.166}$$

The lower bounds for the stability margins developed above depend on the matrices $\{W,Q\}$. In particular, different pairs yield different lower bounds and it is known, see the cited reference, that the bounds which are closest to the actual value of the stability margin are obtained from a pair $\{W,Q\}$ corresponding to the minimum norm of the augmented plant matrix A. Suppose, therefore, that stability along the pass holds and positive definite solutions $\{W,Q\}$ of (4.127) exist. Then the minimum spectral norm, μ, of the corresponding state-space model, or realisation, is defined as

$$\mu = \min_{T} ||T \; AT^{-1}|| \qquad\qquad (4.167)$$

where $||.||$ is any suitable norm, and $T = T_1 \oplus T_2$ where T_1 and T_2 are real constant n×n and m×m matrices respectively.

Summarising, therefore, this section has considered in depth the development of stability tests for example 2.3.8 based on theorem 3.4.4 which shows the equivalence of stability along the pass in this particular case and BIBO stability of 2D linear systems described by the Roesser model. Particular attention has been directed towards a Lyapunov approach and this has yielded two essentially different systematic test procedures and associated tests. The first of these is based in a 2D Lyapunov equation with constant coefficients and the second is based on a 1D Lyapunov equation with coefficients which are functions of a complex variable. Further, the first approach is, in general, sufficient but not necessary but the second is both necessary and sufficient. Detailed comparative studies of these procedures with those of section 4.3 would, however, require the results from applying all of these procedures to suitably defined benchmark problems. Here this wide ranging area has been left for future research with the note that its sufficient, but not necessary, basis will clearly reduce the general usefulness of the first Lyapunov based approach of this section in terms of stability testing.

The application of the constant coefficient Lyapunov equation approach to the problem of developing physically meaningful stability margins for example 2.3.8 has been considered. In particular, some work from the area of 2D linear systems described by the Roesser model has been extended to this case. Further, there are two (interrelated) areas to which future research effort could profitably be directed. These are further development of the basic computational algorithm for increased efficiency, which may necessitate some reformulation of the existing results, and in depth work to establish the correlation (if any) with system performance.

Consider now the first of these two areas. Then progress here will serve to strengthen the already documented links between example 2.3.8 and the Roesser model. In the case of the second area, the final objective here would clearly be to produce easy to use, ideally within a CAD environment, stability and/or performance indicators. One obvious aspect to investigate in this particular case is the links (if any) with the recently introduced concept of a pole for example 2.3.8, defined in terms of the solutions of the two variable polynomial $\rho(z_1, z_2)$ of (4.78), which is the most intuitively appealing definition of a 'characteristic polynomial' for this case. Note, however, that the 'pole concept' for example 2.3.8 (and other cases) is still very much in the development stage and a review of progress to date can be found in the cited reference. The problem of developing stability and/or performance indicators is considered again in the next chapter where alternative simulation-based tests are developed. These then lead to the production, at no extra cost, of computable information concerning the rate of approach to the limit profile in one special case of major practical interest. Finally, chapter 6 considers the use of this information in the formulation of controller design algorithms.

To conclude this section, return to the more general non-unit memory case of example 2.3.7. Then an obvious question to ask is whether or not the analysis of this section generalises in a natural manner and to date no real effort has been directed towards this area. This general question is not considered further here except to note that substantial progress in certain particular aspects should be achieved with relatively little effort. For example, it appears that the second Lyapunov approach should generalise in a straightforward manner and if this is the case then the next stage would be to follow up on the stability margin results.

4.5 Application of Delay Differential Stability Tests

A special case of example 2.3.6 has shown a structural link between differential unit memory linear repetitive processes and a particular sub-class of delay differential systems. The purpose of this section, therefore, is to consider the application of results from this well researched area to example 2.3.4 and, in particular, to the development of stability tests. As a primer to the analysis presented here, the following is a brief summary of the relevant background material. Complete details can, for example, be found in the cited references.

Consider the functional differential equation

$$\frac{d^{n_2}}{dt^{n_2}} Y(t) + \sum_{i=0}^{n_2} \sum_{j=0}^{n_1} c_{ij} \frac{d^i}{dt^i} Y(t - jh) = 0 \qquad (4.168)$$

where the coefficients c_{ij} are real scalars. Then this equation describes a delay differential system with commensurate delays. Further, the following result expresses one form of stability for (4.168) in terms of its so-called characteristic function

$$C(s, e^{-jhs}) := s^{n_2} + \sum_{i=0}^{n_2} \sum_{j=0}^{n_1} c_{ij} s^i e^{-jhs} \qquad (4.169)$$

Theorem 4.5.1: The delay differential system (4.168) is asymptotically stable independent of delay (I.O.D) if, and only if

$$C(s, e^{-jhs}) \neq 0, \quad \forall \ \text{Re}\{s\} \geq 0, \quad h \geq 0 \qquad (4.170)$$

Suppose now that the variable $z := e^{-hs}$, i.e. a left shift operator of duration h, is introduced into (4.170). Then this yields the two-variable polynomial

$$C(s,z) = s^{n_2} + \sum_{i=0}^{n_2} \sum_{j=0}^{n_1} c_{ij} s^i z^j \tag{4.171}$$

which can also be obtained directly from (4.168) by applying the joint (s,z) transform. Given $C(s,z)$, it is always possible to realise (4.168) by the autonomous (no inputs) 2D state-space model

$$\begin{bmatrix} X_1(t+h) \\ \\ \dot{X}_2(t) \end{bmatrix} = \begin{bmatrix} A_1 & A_2 \\ \\ A_3 & A_4 \end{bmatrix} \begin{bmatrix} X_1(t) \\ \\ X_2(t) \end{bmatrix}$$

$$Y(t) = [C_1 \; C_2] \begin{bmatrix} X_1(t) \\ \\ X_2(t) \end{bmatrix} \tag{4.172}$$

where X_1 and X_2 are termed the delayed and continuous state vectors respectively. The characteristic polynomial of this model is

$$C(s,z) = \begin{vmatrix} I_{n_1} - zA_1 & -zA_2 \\ \\ -A_3 & sI_{n_2} - A_4 \end{vmatrix} \tag{4.173}$$

which can be written in the form (4.171). Note also that (4.172) realises both neutral ($c_{n_2,j} \neq 0$ for some $j \in [1,n_1]$) and retarded ($c_{n_2,j} = 0, \; \forall \; j \in [1,n_1]$)

systems. Hence stability conclusions based on this model apply to both cases.

Given (4.172), it is possible to derive sufficient conditions for asymptotic stability of delay differential systems in terms of frequency dependent 1D Lyapunov equations. In particular, define the sets $\bar{D}$ and $\bar{U}$ by

$$\bar{D} = \{s: \; \text{Re}\{s\} \geq 0\} \tag{4.174}$$

and

$$\bar{U} = \{z: \; |z| \leq 1\} \tag{4.175}$$

respectively and form the Cartesian product $\bar{D} \times \bar{U}$. Further, consider the following condition for so-called pointwise asymptotic stability

$$C(s,z) \neq 0 \text{ in } \bar{D} \times \bar{U} \tag{4.176}$$

i.e. the characteristic polynomial is void of zeros in the non-compact biplane composed of the closed right-half plane and the closed unit disc (or circle). Further, (4.176) is a stronger condition than (4.170) since it can be shown that $\bar{D} \times \bar{U}$ has more points than $\bar{D} \times e^{(-\bar{D})}$. Hence pointwise asymptotic stability is more conservative, or stronger, than asymptotic stability (I.O.D). The following result, for which a proof is given since it plays a central role in the analysis which follows, gives necessary and sufficient conditions for (4.176).

<u>Lemma 4.5.1</u>: The characteristic polynomial of the delay differential system described by the 2D model (4.172) satisfies (4.176) if, and only if, the following conditions hold

(a)   all eigenvalues of the matrix A_4 have strictly negative real parts;

and

(b)   all eigenvalues of

$$Z(s) := A_2(sI_{n_2} - A_4)^{-1}A_3 + A_1 \qquad (4.177)$$

with $s = i\omega$ have modulus strictly less than unity for all real frequencies $\omega \geq 0$ or, equivalently,

(c)   all eigenvalues of the matrix A_1 have modulus strictly less than unity;

and

(d)   all eigenvalues of

$$S(z) := A_3(z^{-1}I_{n_1} - A_1)^{-1}A_2 + A_4 \qquad (4.178)$$

have real parts strictly less than zero for all $|z| = 1$.

<u>Proof</u>: By Schur's formula

$$C(s,z) = |I_{n_1} - zA_1||sI_{n_2} - S(z)|$$

$$= |I_{n_1} - zZ(s)||sI_{n_2} - A_4| \qquad (4.179)$$

Using (4.179), the rest of the proof is a straightforward application of the maximum modulus theorem. ■

Return now to the state-space model of example 2.3.4 and delete the current pass input terms to yield (with $Y_{k+1}(0) = Y_k(\alpha)$) the autonomous version

$$\dot{X}_{k+1}(t) = AX_{k+1}(t) + B_0Y_k(t)$$
$$Y_{k+1}(t) = CX_{k+1}(t) + D_1Y_k(t)$$
$$0 \leq t \leq \alpha, \quad k \geq 0 \qquad (4.180)$$

Then it follows immediately that (4.180) can be modelled by the following special case of the 2D state-space model of (4.172)

$$\begin{bmatrix} \dot{X}(t) \\ \\ Y(t+\alpha) \end{bmatrix} = \begin{bmatrix} A & B_0 \\ \\ C & D_1 \end{bmatrix} \begin{bmatrix} X(t) \\ \\ Y(t) \end{bmatrix} \qquad (4.181)$$

where here $X(t)$ denotes the current pass state vector $X_{k+1}(t)$ and $Y(t)$ denotes the previous pass profile $Y_k(t)$. Further, the following result now shows that an equivalence exists between stability along the pass of example 2.3.4 and pointwise asymptotic stability of its delay differential interpretation.

<u>Theorem 4.5.2</u>: Regard the model of example 2.3.4 (in its autonomous form) as a delay differential system described by the 2D state-space model of (4.181). Then the extended linear repetitive process $S(E_\alpha, W_\alpha, L_\alpha)_{\alpha \geq \alpha_0}$ generated by this model with $\alpha \geq \alpha_0$ is stable along the pass if, and only if, it is pointwise asymptotically stable in the sense of (4.176).

<u>Proof</u>: This, in effect, consists of showing that the conditions of lemma 4.5.1 (in particular, the set consisting of (a) and (b)) and corollary 3.3.10 are equivalent.

Consider first, therefore, the delay differential interpretation of (4.181). Then (a) and (b) of lemma 4.5.1 translate to the following conditions for pointwise asymptotic stability:

(a) all eigenvalues of the matrix A have strictly negative real parts; and

(b) all eigenvalues of

$$G_1(s) = C(sI_n - A)^{-1}B_0 + D_1 \qquad (4.182)$$

with $s = i\omega$ have modulus strictly less than unity for all real frequencies $\omega \geq 0$.

Further, it follows immediately that all eigenvalues of D_1 must have modulus strictly less than unity (i.e. asymptotic stability) in order for (b) to hold. Hence these two conditions are equivalent to (a) - (c) of corollary 3.3.10 for stability along the pass.

Conversely, suppose that (a) and (b) above hold. Then the proof that these imply stability along the pass is identical to that of corollary 3.3.10 and is hence omitted.   ∎

Theorem 4.5.2 can be regarded as the analogous result to theorem 3.4.4 which established the equivalence between stability along the pass of the discrete unit memory process of example 2.3.8 and BIBO stability of 2D linear systems described by the Roesser model of (2.58). The following analysis mirrors section 4.4 for the discrete case in considering the use of theorem 4.5.2 as the basis for the development of stability tests to serve as alternatives, and/or supplements to, those of section 4.2.

A study of the published literature on the development of stability tests for delay differential systems shows that this problem has been studied from a variety of starting points. One major approach has centred on applying a root clustering based argument to $C(s,z)$ of (4.176) and uses techniques such as the Schur-Cohn matrix and modified Routh array, which were also used in section 4.2 to develop tests for stability along the pass of the differential non-unit memory process of example 2.3.3. Consequently the analysis below concentrates on a Lyapunov approach and complete details of the various other approaches can be found in the cited references.

In a similar manner to that for 2D linear systems described by the Roesser model, the general problem of developing a Lyapunov approach to the stability analysis of

delay differential systems, and hence example 2.3.4 by theorem 4.5.2, has been
studied in two (essentially different) ways. One of these consists of developing a
1D Lyapunov equation with coefficients which are functions of a complex parameter
and the other is based on a 2D Lyapunov equation with constant coefficients. In
either case, the basic objective here is a suitable extension of the Lyapunov theory
for the derived system $L_D(A,B,C,D_0)$ which gives necessary and sufficient conditions
for stability along the pass of example 2.3.4.

Considering first the 1D Lyapunov equation approach yields the following dual
approach to that of theorem 4.4.2, and hence theorem 4.4.3, for the discrete unit
memory process of example 2.3.8. Hence the result is stated without proof.

<u>Theorem 4.5.3:</u> The extended linear repetitive process $S(E_\alpha,W_\alpha,L_\alpha)_{\alpha \geq \alpha_0}$ generated by

the model of example 2.3.4 with $\alpha \geq \alpha_0$ is stable along the pass if, and only if,

(a) all eigenvalues of the matrix A have strictly negative real parts or,
     equivalently, the derived conventional linear system $L_D(A,B,C,D_0)$ is stable;
     and

(b) the matrix Lyapunov equation
$$H(s) - G_1^*(s)H(s)G_1(s) = W(s) \qquad\qquad (4.183)$$

has a unique P.D.H. solution $H(s)$ for any P.D.H. matrix $W(s)$, $s = i\omega$ for any
$\omega \geq 0$.   ∎

Note: As in section 4.4, P.D.H. denotes positive definite Hermitian and * denotes
the complex conjugate transpose, i.e. $H^*(i\omega) = H^T(-i\omega)$.

Recall now the unit memory versions of the systematic test procedures developed
in section 4.2. Then, in effect, theorem 4.5.3 serves as the basis for an
alternative to either of these. Further, these new conditions should obviously be
tested in the order of (a) followed by (b) with termination if the former does not
hold.

Theorem 4.5.3 is the differential equivalent of theorem 4.4.3 for the discrete
process of example 2.3.8. Further, the testing of (a) in this case is just the
standard stability problem for $L_D(A,B,C,D_0)$. Tests for (b) can also be developed by
following analogous steps to those used in section 4.4 for (c) of theorem 4.4.3.
Consequently these are not detailed here and complete details can be found in the
cited references. Instead, the following analysis gives an introduction as to how
the Lyapunov equation (4.183) can be used to provide a physically based
interpretation of stability along the pass in this particular case. Complete
details, including the corresponding analysis for example 2.3.8, can again be found
in the cited reference.

Noting again the causality definition of (2.59) and Figure 2.7, apply the Laplace
transform to the autonomous model of (4.181) to yield

$$Y_{k+1}(s) = G_1(s)Y_k(s) , \quad k \geq 0 \tag{4.184}$$

Further, suppose that the output $Y_{k+1}(s)$ is passed through a filter with transfer-function matrix $R(s)$ and denote the result of this operation by $\hat{Y}_{k+1}(s)$, $k \geq 0$, i.e.

$$\hat{Y}_{k+1}(s) = R(s)Y_{k+1}(s) = R(s)G_1(s)Y_k(s), \quad k \geq 0 \tag{4.185}$$

Then, by Parseval's theorem,

$$\int_0^\infty \hat{Y}_{k+1}^T(t)\hat{Y}_{k+1}(t)dt = \frac{1}{2\pi} \int_0^\infty \hat{Y}_{k+1}^*(i\omega)\hat{Y}_{k+1}(i\omega)d\omega \tag{4.186}$$

$$= \frac{1}{2\pi} \int_0^\infty Y_k^*(i\omega)G_1^*(i\omega)R^*(i\omega)R(i\omega)G_1(i\omega)Y_k(i\omega)d\omega \tag{4.187}$$

Suppose also that $H(s) = R^*(s)R(s)$ satisfies the Lyapunov equation (4.183). In which case

$$\int_0^\infty \hat{Y}_{k+1}^T(t)\hat{Y}_{k+1}(t)dt = \frac{1}{2\pi} \int_0^\infty Y_k^*(i\omega)(H(i\omega) - W(i\omega))Y_k(i\omega)d\omega$$

$$= \frac{1}{2\pi} \int_0^\infty \hat{Y}_k^*(i\omega)\hat{Y}_k(i\omega)d\omega - \frac{1}{2\pi} \int_0^\infty Y_k^*(i\omega)W(i\omega)Y_k(i\omega)d\omega \tag{4.188}$$

and hence

$$\int_0^\infty \hat{Y}_{k+1}^T(t)\hat{Y}_{k+1}(t)dt < \int_0^\infty \hat{Y}_k^T(t)\hat{Y}_k(t)dt, \quad k \geq 0 \tag{4.189}$$

Equivalently, the filtered output, $\hat{Y}$, decreases in amplitude from pass to pass in an L_2 sense.

In the case of delay differential systems, the so-called continuous bounded real and discrete positive real lemmas from circuit theory have been used to develop 2D Lyapunov equations for the stability conditions of lemma 4.5.1 and, in particular, (4.177) and (4.178) respectively. As shown below, the repetitive systems case requires only the former concept strengthened to so-called strictly continuous bounded real which is denoted by S.C.B.R. Further, let $W = W_1 \oplus W_2$ and Q be real symmetric positive definite matrices. Then the proposed 2D Lyapunov equation for example 2.3.4 is

$$P^T W^{1,0} + W^{1,0}P + P^T W^{0,1}P - W^{0,1} = -Q \tag{4.190}$$

where $W^{1,0} := W_1 \oplus 0_m$, $W^{0,1} := 0_n \oplus W_2$ and P is the so-called augmented plant matrix defined as

$$P = \begin{bmatrix} A & B_0 \\ C & D_1 \end{bmatrix} \tag{4.191}$$

The first step in relating (4.191) to stability along the pass is the following formal definition of the term S.C.B.R.

<u>Definition 4.5.1</u>: Consider a square matrix $Z(s)$ over $R(s)$, the ring of polynomials in s over the real line. Then $Z(s)$ is termed S.C.B.R. if the following conditions hold:

(i) $Z(s)$ is analytic in $\bar{D}$; and

(ii) $I - Z^*(s)Z(s) > 0$, $s = i\omega, \forall$ real $\omega \geq 0$ (4.192)
■

Further, interpreting this definition in terms of the interpass transfer-function matrix $G_1(s)$ and comparing the resulting conditions with those of theorem 4.5.3 immediately shows that (ii) here is equivalent to (4.183) admitting the constant solution $H = T^*T$ over the real line. Hence it follows immediately that S.C.B.R. implies stability along the pass. In general, however, the converse is not true, a result which parallels that of section 4.4 for the 2D Lyapunov equations introduced there for the discrete process of example 2.3.8. This result is proved by straightforward modifications to the steps used in establishing its discrete counterpart and hence the details are omitted.

Despite its sufficient but not necessary basis, which clearly reduces its usefulness in terms of stability tests given the necessary and sufficient alternatives of section 4.2, the 2D Lyapunov equation still has a (potentially) significant role to play in certain aspects of the analysis of example 2.3.4. One of these is to develop a first attempt at the wide ranging problem of constructing useful, and 'easy to use', stability margins. This topic is returned to later in this section after stability along the pass has been formally expressed in terms of (4.190).

Suppose, therefore, that the quadruple $\{F,G,J^T,K\}$ is a minimal realisation of $Z(s)$, i.e.
$$Z(s) = J^T(sI - F)^{-1}G + K \qquad\qquad (4.193)$$
Then the following lemma can be introduced and leads directly to the required result in the form of theorem 4.5.4 below. This so-called S.C.B.R. lemma is a well known result in circuit theory and its proof can, for example, be found in the cited reference.

<u>Lemma 4.5.2</u>: Suppose that the quadruple $\{F,G,J^T,K\}$ is a minimal realisation of $Z(s)$ of (4.193). Then $Z(s)$ is S.C.B.R. if, and only if, there exists a positive definite symmetric matrix P_1 such that

$$Q_1 := \begin{bmatrix} F^T P_1 + P_1 F + JJ^T & (P_1 G + JK) \\ (P_1 G + JK)^T & K^T K - I \end{bmatrix} < 0 \qquad\qquad (4.194)$$

i.e. Q_1 is negative definite symmetric.

                                                                           ■

Theorem 4.5.4: Consider the differential unit memory linear repetitive process of example 2.3.4 under the controllability and observability assumptions of corollary 3.3.8, i.e. the pair $\{A, B_0\}$ is controllable and the pair $\{C, A\}$ is observable.

Suppose also that there exists a matrix T, non-singular over the real line, such that

$$\tilde{G}_1(s) = T[C(sI_n - A)^{-1}B_0 + D_1]T^{-1} \tag{4.195}$$

is S.C.B.R. Then there exists positive definite symmetric matrices W_1, W_2 and Q such that $W = W_1 \oplus W_2$ and Q satisfy the 2D Lyapunov equation (4.190). Conversely, suppose that (4.190) holds for positive definite symmetric matrices $W = W_1 \oplus W_2$ and Q, then there exists a matrix T, non-singular over the real line, such that $\tilde{G}_1(s)$ is S.C.B.R.

Proof: Consider first sufficiency and note that $G_1(s)$ has the minimal realisation $\{A, B_0 T^{-1}, TC, TD_1 T^{-1}\}$ and is S.C.B.R. Hence by lemma 4.5.2 there exists a positive definite symmetric matrix P_1 such that

$$Q_2 := \begin{bmatrix} A^T P_1 + P_1 A + C^T T^T TC & P_1 B_0 T^{-1} + C^T T^T TD_1 T^{-1} \\ (P_1 B_0 T^{-1} + C^T T^T TD_1 T^{-1})^T & (T^{-1})^T D_1^T T^T TD_1 T^{-1} - I_m \end{bmatrix} < 0 \tag{4.196}$$

and this yields, after pre-multiplication by $(I_n \oplus T^T)$ and post-multiplication by $(I_n \oplus T)$, the 2D Lyapunov equation (4.190) with

$$Q = -(I_n \oplus T^T)Q_2(I_n \oplus T) > 0 \tag{4.197}$$

and

$$W = (P_1 \oplus T^T T) > 0 \tag{4.198}$$

To prove necessity, suppose that (4.190) holds and set $T = W_2^{\frac{1}{2}}$, $F = A$, $G = B_0 T^{-1}$, $J^T = TC$ and $K = TD_1 T^{-1}$. Then pre and post-multiply (4.190) by $(I_n \oplus (T^{-1})^T)$ and $(I_n \oplus T^{-1})$ respectively to yield (4.194) with $P_1 = W_1$ and $Q_1 = -(I_n \oplus (T^{-1})^T) Q(I_n \oplus T^{-1})$. Hence Z(s) is S.C.B.R. as required. ∎

As noted previously, S.C.B.R. is a special case of the 2D Lyapunov equation (4.183) and hence the following corollary can be stated.

Corollary 4.5.4: Consider the extended linear repetitive process $S(E_\alpha, W_\alpha, L_\alpha)_{\alpha \geq \alpha_0}$ generated by the model of exmaple 2.3.4 with $\alpha \geq \alpha_0$. Then theorem 4.5.3 for stability along the pass holds if there exists positive definite symmetric matrices $W = W_1 \oplus W_2$ and Q such that the 2D Lyapunov equation (4.190) holds.

In common with the 2D Lyapunov equation used in section 4.4 for the discrete unit memory process, a number of special cases exist where (4.190) yields necessary and sufficient conditions. One of these is the SISO case where it is easily shown that corollary 3.3.10 holds if, and only if, the interpass transfer-function $G_1(s)$ is S.C.B.R. This result is stated formally in the following corollary.

<u>Corollary 4.5.5:</u> The extended linear repetitive process $S(E_\alpha, W_\alpha, L_\alpha)_{\alpha \geq \alpha_0}$ generated by the model of example 2.3.4 with $m = 1$ and $\alpha \geq \alpha_0$ is stable along the pass if, and only if, the interpass transfer-function $G_1(s)$ is S.C.B.R.
∎

Work is proceeding on the development of efficient algorithms for computing the matrix P_1 (if it exists). Further, other conditions for S.C.B.R. have been developed using, for example, algebraic Riccati equations. Complete details of progress to date can be found in the cited references.

The remainder of this section uses the 2D Lyapunov equation (4.190) as a basis for the first attempt at developing useful, and 'easy to use', stability margins for example 2.3.4. In particular, it follows the approach of section 4.4 and extends some work from delay differential systems to this case. Further, the potential for other developments in this area is briefly discussed.

Suppose, therefore, that there exists positive definite symmetric matrices Q, W_1, W_2 and $W = W_1 \oplus W_2$ such that (4.190) holds. Then this is, in general, a sufficient condition for stability along the pass and hence

$$\rho(s,z) = \begin{vmatrix} sI_n - A & - B_o \\ - zC & I_m - zD_1 \end{vmatrix} \tag{4.199}$$

satisfies

$$\rho(s,z) \neq 0 \text{ in } \bar{D} \times \bar{U} \tag{4.200}$$

i.e. the stability region excludes the non-compact biplane composed of the closed right-half plane and the closed unit disc (or circle). Further, as for the discrete process of example 2.3.8 considered in section 4.4, the stability margins are defined (in common with the delay differential case) as the shortest distances between the roots of $\rho(s,z)$ and the boundaries of $\bar{D} \times \bar{U}$. In particular, given stability along the pass, these margins are defined as the largest values of the scalars δ and σ for which

$$\rho(s,z) \neq 0 \text{ in } \bar{D} \times \bar{U}_\delta \tag{4.201}$$

and

$$\rho(s,z) \neq 0 \text{ in } \bar{D}_\sigma \times \bar{U} \tag{4.202}$$

respectively where

$$\bar{U}_\delta = \{z: |z| \leq 1 + \delta\} \tag{4.203}$$

and

$$\bar{D}_\sigma = \{s: \ \mathrm{Re}\{s\} \geq -\sigma\} \tag{4.204}$$

Due to the non-compactness of $\bar{D} \times \bar{U}$, a combined stability margin (corresponding to σ of (4.131) in section 4.4 for example 2.3.8) is not defined in this case since it has no clearly defined meaning.

Consider now the case when positive definite solutions W and Q exist for the Lyapunov equation (4.190) where $W = W_1 \oplus W_2$ and

$$Q = \begin{bmatrix} Q_1 & Q_2^T \\ Q_2 & Q_3 \end{bmatrix} \tag{4.205}$$

Then this is, in general, a sufficient condition for stability along the pass and hence (4.200) holds. Further, the following analysis yields lower bounds for δ and σ in terms of W_1, W_2, Q_1, Q_2 and Q_3. First pre and post-multiply (4.190) by $(I_n \oplus \beta I_m)$, where β is a real scalar, to obtain

$$\begin{bmatrix} A & \beta B_0 \\ C & \beta D_1 \end{bmatrix}^T \begin{bmatrix} W_1 & 0 \\ 0 & 0 \end{bmatrix} \begin{bmatrix} W_1 & 0 \\ 0 & 0 \end{bmatrix} \begin{bmatrix} A & \beta B_0 \\ C & \beta D_1 \end{bmatrix} +$$

$$\begin{bmatrix} A & \beta B_0 \\ C & \beta D_1 \end{bmatrix}^T \begin{bmatrix} 0 & 0 \\ 0 & W_2 \end{bmatrix} \begin{bmatrix} A & \beta B_0 \\ C & \beta D_1 \end{bmatrix} = - \begin{bmatrix} Q_1 & \beta Q_2^T \\ \beta Q_2 & \beta^2 (Q_3 - W_2) \end{bmatrix} \tag{4.206}$$

Subtracting $(2\gamma W_1 \oplus W_2)$, where γ is also a real scalar, from both sides of (4.206) now yields the following equation after some algebraic manipulations

$$\hat{P}^T W^{1,0} + W^{1,0} \hat{P} + \hat{P}^T W^{0,1} \hat{P} - W^{0,1} = -\hat{Q} \tag{4.207}$$

where

$$\hat{P} = \begin{bmatrix} A - \gamma I_n & \beta B_0 \\ C & \beta D_1 \end{bmatrix} \tag{4.208}$$

and

$$\hat{Q} = \begin{bmatrix} Q_1 + 2\gamma W_1 & \beta Q_2^T \\ \beta Q_2 & \beta^2 (Q_3 - W_2) + W_2 \end{bmatrix} \tag{4.209}$$

Suppose also that $\hat{Q}$ is positive definite and hence, since W is assumed positive definite, a sufficient condition holds for

$$\hat{\rho}(s,z) \neq 0 \text{ in } \bar{D} \times \bar{U} \tag{4.210}$$

where

$$\hat{\rho}(s,z) = \begin{vmatrix} sI_n - (A - \gamma I_n) & -\beta B_o \\ -zC & I_m - z\beta D_1 \end{vmatrix}$$

(4.211)

The following result, whose proof follows similar steps to that of lemma 4.4.1 for the discrete process of example 2.3.8 and is hence omitted, now establishes the relationship between $\rho(s,z)$ and $\hat{\rho}(s,z)$.

Lemma 4.5.3: Consider $\rho(s,z)$ and $\hat{\rho}(s,z)$ defined by (4.199) and (4.211) respectively. Then

$$\rho(s,z) = \hat{\rho}(s - \gamma, \beta^{-1}z)$$

(4.212)

Using this result (compare with lemma 4.4.1 for the discrete case), it is ■ possible to characterise the roots of $\hat{\rho}(s,z)$ as functions of γ and β. In particular, if $\gamma = 0$ and $\beta = 1$ then it is obvious that $\rho(s,z)$ and $\hat{\rho}(s,z)$ are identical and satisfy (4.200). If, however, $\beta > 1$ then, for $\gamma > 0$, the roots of $\hat{\rho}(s,z)$ move from those of $\rho(s,z)$ towards the boundary of $\bar{D} \times \bar{U}$ and ultimately cross it. Further, note again that $\hat{Q} > 0$ is sufficient for $\hat{\rho}(s,z)$ to satisfy (4.210) and hence the range of α and β for which this matrix remains positive definite can provide limits for the stability margins defined by (4.201) - (4.202). Since, however, this is, in general, a sufficient, but not necessary condition it follows immediately that the range of γ and β for which $\hat{Q}$ is positive definite can only give lower bounds, not actual values, for these margins. The analysis below obtains such bounds.

Considering first δ, it is clear that a lower bound in this case can be obtained from the range of $\beta > 1$ for which

$$\hat{Q} = \begin{bmatrix} Q_1 & \beta Q_2^T \\ \beta Q_2 & \beta^2(Q_3 - W_2) + W_2 \end{bmatrix} > 0$$

(4.213)

This is equivalent to

$$Q_1 > 0$$

(4.214)

and

$$W_2 - \beta^2(W_2 - Q_3 + Q_2 Q_1^{-1} Q_2^T) > 0$$

(4.215)

Further, (4.214) holds by assumption and hence it remains to consider (4.215) which is equivalent to

$$(W_2 - Q_3 + Q_2 Q_1^{-1} Q_2^T) - \beta^{-2} W_2 < 0$$

(4.216)

where this new condition is expressed in terms of a matrix pencil of the form $F - \lambda B$, with $\lambda = \beta^{-2}$

$$B = W_2$$

(4.217)

and

$$F = W_2 - Q_3 + Q_2 Q_1^{-1} Q_2^T \tag{4.218}$$

Note also that this pencil is regular since $W_2 > 0$ and F can be shown to be positive semi-definite on observing from (4.190) that

$$W_2 - Q_3 = D_1^T W_2 D_1 \geq 0 \tag{4.219}$$

The extremal properties of the characteristic values (eigenvalues) of the pencil $F - \lambda B$ can now be examined using well established theory. In particular, the result listed under (4.143) in section 4.4 is relevant, i.e.

$$F - \lambda_{max}[B^{-1}F]B \leq 0 \tag{4.220}$$

where $\lambda_{max}[B^{-1}F]$ again denotes the maximum eigenvalue of $B^{-1}F$. Noting again that $B = W_2$ is positive definite by assumption and F of (4.218) is positive semi-definite, (4.220) now yields the values of β for which (4.215) holds as

$$\beta^{-1} > \sqrt{\lambda}_{max}[B^{-1}F] \tag{4.221}$$

or

$$\beta < (\sqrt{\lambda}_{max}(I_m - W_2^{-1}Q_3 + W_2^{-1}Q_2 Q_1^{-1}Q_2^T))^{-1} \tag{4.222}$$

Further, it is easily shown that all eigenvalues of $B^{-1}F$ are real and non-negative. Hence a lower bound for δ can be written as

$$\delta \geq (\sqrt{\lambda}_{max}(I_m - W_2^{-1}Q_3 + W_2^{-1}Q_2 Q_1^{-1}Q_2^T))^{-1} - 1 \tag{4.223}$$

In a similar manner to the case of δ above, a lower bound for σ can be obtained from the range of $\gamma > 0$ for which

$$\hat{Q} = \begin{bmatrix} Q_1 + 2\gamma W_1 & Q_2^T \\ Q_2 & Q_3 \end{bmatrix} > 0 \tag{4.224}$$

Further, since $Q_3 > 0$ by assumption, this is equivalent to

$$Q_1 - Q_2^T Q_3^{-1} Q_2 + 2\gamma W_1 > 0 \tag{4.225}$$

This is again a regular matrix pencil of the form $F - \lambda B$, with $\lambda = -\gamma$

$$B = 2W_1 > 0 \tag{4.226}$$

and

$$F = Q_1 - Q_2^T Q_3^{-1} Q_2 \tag{4.227}$$

Using (4.190), it is easily shown that $F > 0$ and

$$F - \lambda_{min}[B^{-1}F]B \geq 0 \tag{4.228}$$

is a well known result from established theory where $\lambda_{min}[B^{-1}F]$ denotes the minimum eigenvalue of $B^{-1}F$. These facts now yield a lower bound for σ as

$$\sigma \geq \frac{1}{2} \lambda_{min}[W_1^{-1}Q_1 - W_1^{-1}Q_2^T Q_3^{-1} Q_2] \tag{4.229}$$

The stability margin bounds δ and σ developed above have been derived assuming the availability of positive definite solutions for the 2D Lyapunov equation (4.190). As for the discrete case of section 4.4, such solutions may not exist but, if they do, recently reported work, see the cited references for complete details, strongly suggests that they can be computed using similar algorithms to those for the discrete case. This should then lead directly to efficient algorithms for computing the stability margin bounds derived here.

Summarising, therefore, this section has considered in depth the application of results from the stability theory of delay differential systems to example 2.3.4. The central result is theorem 4.5.2 which shows that stability along the pass is equivalent to pointwise asymptotic stability when this example is interpreted as a delay differential system. This result has then been used to develop a Lyapunov approach to stability testing which has yielded two essentially different approaches. In particular, tests based on a 1D Lyapunov equation with coefficients which are functions of a complex variable and a 2D Lyapunov equation with constant coefficients have been developed. Further, as in the discrete case of section 4.4, the first approach is necessary and sufficient but the second is, in general, sufficient only. Detailed comparative studies with the systematic test procedures of section 4.2 would, however, require the results from application of all of these tests to suitably defined benchmark problems. Here this wide ranging area has been left for future research with the note that its sufficient, but not necessary, basis will clearly reduce the general usefulness of the 2D Lyapunov equation approach in this context.

The application of the 2D Lyapunov equation approach to the problem of developing physically meaningful stability margins for example 2.3.4 has been considered. In particular, some work from the delay differential systems area has been extended to this case. Further, there are two (interrelated) areas to which future research effort could profitably be directed. These are the devleopment of efficient computational algorithms and in depth work to establish the correlation (if any) with system performance.

Progress on the first of these areas will serve to further strengthen the already documented links between example 2.3.4 and certain classes of delay differential systems. In the case of the second, the final objective (as in the corresponding case for the discrete process of example 2.3.8) would clearly be to produce 'easy to use', ideally within a CAD environment, stability and/or performance indicators. One obvious aspect to investigate in this particular case is the links (if any) with the recently introduced concept of a pole for example 2.3.4, defined in terms of solutions of the two variable polynomial $\rho(s,z)$ of (4.199), which is the most intuitively appealing definition of a characteristic polynomial for this case. (See also section 6.2). (Again compare with the discrete case of section 4.4). Note also that the problem of developing stability and/or performance indicators is considered again in the next chapter using the alternative simulation-based tests developed there. These tests lead (as in the discrete case) to the production, at no extra cost, of computable information concerning the rate of approach to the limit profile in one special case of major practical interest. Finally, chapter 6 considers the use of this information in the formulation of controller design algorithms.

To conclude this section, return to the more general non-unit memory case of example 2.3.3. Then an obvious question to ask is whether or not the analysis of this section generalises in a natural manner. Some promising preliminary results in this area can be found in the cited reference.

Notes and References

The stability tests of sections 4.2 and 4.3 are due to Rogers and Owens (1989a,b,c, 1990a) and make use of the theory of axis and circle positivity due to Siljak (1971,1973,1975). Complete details of both the conventional linear systems tests used here and the modified Routh array can be found in Jury (1974), Gantmacher (1959) and Siljak (1971). Background on the characteristic locus can be found in Postlethwaite and MacFarlane (1979).

Section 4.4 has evolved from the work of Boland and Owens (1980) as detailed in section 3.4. The background on the standard Lyapunov theory can, for example, be found in Willems (1970) and that on the nD version is from Piekarski (1977). Anderson, Agathoklis, Jury and Mansour (1986) is the basis for the 2D Lyapunov equation, the special case of lemma 4.4.2 is from the same source and that of theorem 4.4.1 is based on Fadali and Gnanasekaran (1989). Algorithms for computing solutions to the 2D Lyapunov equation can be found in Agathoklis, Jury and Mansour (1989). Theorems 4.4.2 and 4.4.3, the essence of the 1D Lyapunov equation approach are based on Lu and Lee (1985) as is the special case of corollary 4.4.3. Theorems 4.4.4 and 4.4.5 are based on Agathoklis, Jury and Mansour (1990) and the background Kronecker product results can, for example, be found in Lancaster and Tismenetsky (1985). Agathoklis (1988) is the basis for the definitions of, and lower bounds for, the stability margins. Rogers and Owens (1990b) contains preliminary results on poles for this case, the role of the stability margins in classifying system performance, and the extensions of these concepts to the non-unit memory case.

Complete details of the delay differential stability results which form the basis of section 4.5 can be found in Hale (1977), Kamen (1982) and Agathoklis and Foda (1989a,b). The S.C.B.R. lemma is based on the work in Anderson and Vongpanitlerd (1973) and work on the use of algebraic Riccati equations in this context can be found in Gu and Lee (1989). Agathoklis and Foda (1989b) forms the basis for the definitions of, and lower bounds for, the stability margins. Rogers and Owens

(1990c) contains the details of the use of the 1D Lyapunov equation to provide a physically based interpretation of stability along the pass. Details of the work to date on poles for this case, the role of the stability margins in classifying system performance, and the extensions of these concepts to the non-unit memory case can be found in Rogers and Owens (1990b).

CHAPTER 5

SIMULATION-BASED STABILITY TESTS

In this chapter simulation-based tests for stability along the pass of the differential and discrete processes of examples 2.3.3 and 2.3.7 are developed. The starting point for these tests is the assumption that suitably well behaved plant step response data is available or can be obtained by simulation studies. Further, it is shown that these tests produce, at no extra cost, computable information concerning the rate of approach to the limit profile, together with bounds on the performance along any pass, in one special case of major practical interest. This information is unique to these tests and its use in the formulation of controller design algorithms is considered in the next chapter. Finally, some initial results on extending these tests to processes with interpass smoothing (see example 2.3.5) are presented.

5.1 Mathematical Background

This section reviews the mathematical background necessary to develop the basic results of the next two sections. Complete details in all cases can be found in the cited references and the content of this section begins with the following result.

Lemma 5.1.1: Suppose that $g \in L_1(0,T)$, d is a real scalar and

$$f(t) := d + \int_0^t g(t')dt' \tag{5.1}$$

is bounded and continuous on the infinite open interval $0 < t < +\infty$ with local maxima and minima at times $t_1 < t_2 < \ldots$ satisfying $\sup t_j = +\infty$ in the extended half-line $t > 0$. Then, with $t_0 = 0$,

$$N_T(f) = |d| + \int_0^T |g(t)|dt \tag{5.2}$$

where

$$N_T(f) := |f(0+)| + \sum_{k=1}^{k^*} |f(t_k) - f(t_{k-1})| + |f(T) - f(t_{k^*})| \tag{5.3}$$

and

$$N_\infty(f) := \sup_{T \geq 0} N_T(f) \tag{5.4}$$

where k^* is the largest integer k such that $t_k < T$.

Proof: Local maxima and minima of f correspond to points where g changes sign. The result now follows immediately on writing

$$|d| + \int_0^T |g(t)|dt = |d| + \sum_{k=1}^{k^*} \int_{t_{k-1}}^{t_k} |g(t)|dt + \int_{t_{k^*}}^T |g(t)|dt$$

$$= |d| + \sum_{k=1}^{k^*} |\int_{t_{k-1}}^{t_k} g(t)dt| + |\int_{t_{k^*}}^{T} g(t)dt| \qquad (5.5)$$

and noting that $f(0+) = d$ and

$$\int_a^b g(t)dt = g(b) - g(a) \qquad (5.6)$$

for any $b \geq a \geq 0$. ∎

The quantity $N_T(f)$ is simply the norm of f regarded as a function of bounded variation on the half-open interval $0 < t \leq T$. Hence it is termed the total variation of f. For each f, $N_T(f)$ is monotonically increasing and continuous as a function of T and therefore $N_\infty(f)$ can be obtained as

$$N_\infty(f) = \lim_{T \to +\infty} N_T(f) \qquad (5.7)$$

Further, $N_T(f)$ is easily computed from simple graphical operations on $f(t)$ as illustrated in Figure 5.1. These operations are easily included in a CAD environment and the cited reference gives further details on this aspect. Note also that

$$\lim_{T \to +\infty} |N_\infty(f) - N_T(f)| = 0 \qquad (5.8)$$

and consequently $N_\infty(f)$ can be accurately estimated using data on a 'long enough' time interval $0 < t \leq T$. On such an interval, the continuity of $N_T(f)$ as a function of the stationary points $t_1, t_2, \ldots$ implies that it is insensitive to errors in their estimation. If $f(t)$ is contaminated by noise $n(t)$, $N_T(f)$ must be evaluated by inspection of $f(t) + n(t)$ where, under the assumption of a 'sufficiently large' signal to noise ratio, the stationary points of $f(t)$ can be estimated to 'reasonable' accuracy by visual smoothing of the recorded response $f(t) + n(t)$. This, together with (5.8), leads to the conclusion that the estimation of $N_\infty(f)$ is a 'robust operation' in many practical situations.

In the discrete case, the following is the equivalent result to lemma 5.1.1. It is stated here without proof since, in effect, this follows identical steps to that of lemma 5.1.1 but without the complications introduced by continuity.

Lemma 5.1.2: Suppose that the sequence defined by $f(0) = d$ and

$$f(r) = d + \sum_{j=1}^{r} g(j), \quad r \geq 1 \qquad (5.9)$$

is bounded with local maxima or minima at sample instants $1 \leq r_1 < r_2 < \ldots$ in the extended positive integers. Then, with $r_0 = 0$,

$$N_r(f) = |d| + \sum_{j=1}^{r} |g(j)| \qquad (5.10)$$

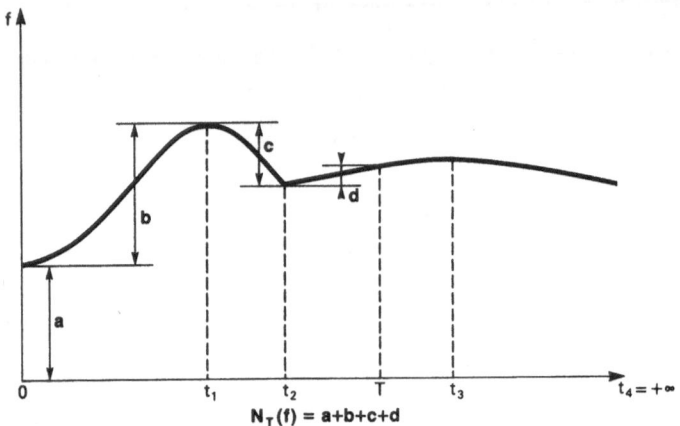

$$N_T(f) = a+b+c+d$$

FIGURE 5.1

where

$$N_r(f) = |f(0)| + \sum_{j=1}^{r^*(r)} |f(r_j) - f(r_{j-1})|$$
$$+ |f(r) - f(r^*(r))| \qquad (5.11)$$

and

$$N_\infty(f) := \sup_{r \geq 0} N_r(f) \qquad (5.12)$$

where $r^*(r)$ is the largest integer $r_1, r_2, \ldots$ satisfying $r_j < r$. ∎

The cited reference again gives further information on the evaluation of $N_r(f)$ for a given sequence $f(r)$.

An essential underlying element of the analysis presented in the next two sections will be the following definitions and results from the theory of non-negative matrices. The proofs of the results listed are well known in the study of such matrices and are hence omitted.

<u>Definition 5.1.1</u>: The partial ordering $\leq$ on $n_1 \times n_2$ matrices is defined by the relation

$$A \leq B \qquad (5.13)$$

if, and only if,

$$A_{ij} \leq B_{ij}, \ \forall i,j \qquad (5.14)$$

Further, the 'absolute value' of an $n_1 \times n_2$ matrix A is defined to be the $n_1 \times n_2$ real, or so-called non-negative, matrix

$$||A||_p = \begin{bmatrix} |A_{11}| & & |A_{1n_2}| \\ & & \\ |A_{n_11}| & & |A_{n_1n_2}| \end{bmatrix} \qquad (5.15)$$

<u>Lemma 5.1.3</u>: The absolute value, $||A||_p$, of an $n_1 \times n_2$ matrix A has the following 'norm like' properties

(i) $\quad ||A||_p \geq 0$ $\qquad (5.16)$

(ii) $\quad ||\gamma A||_p = |\gamma| ||A||_p$, for all complex scalars γ $\qquad (5.17)$

(iii) If B is another $n_1 \times n_2$ matrix then

$$||A + B||_p \leq ||A||_p + ||B||_p \qquad (5.18)$$

(iv) If B is another matrix compatible for pre-multiplication by A then

$$||AB||_p \leq ||A||_p ||B||_p \qquad (5.19)$$

(v) If A and B are square matrices then

$$0 \leq ||A||_p \leq B \Rightarrow r(A) \leq r(||A||_p) \leq r(B) \qquad (5.20)$$

where $r(.)$ again denotes the spectral radius.

∎

Lemma 5.1.4: If A is an $n_1 \times n_1$ matrix then $(I_{n_1} - ||A||_p)^{-1}$ exists and is

non-negative if, and only if,

$$r(||A||_p) < 1 \tag{5.21}$$

∎

The following definitions and results summarise the essential required background theory from functional analysis. Again the proofs required are standard and hence omitted.

Definition 5.1.2: Let X be a Banach space (subsequently specialised to $X = L_\infty(0,+\infty)$) and X^d its dth Cartesian product regarded as the linear vector space of columns $X = (x_1, x_2, \ldots, x_d)^T$ of elements of X. Then the absolute value of $x \in X^d$ is defined by

$$||x||_p = (||x_1||, ||x_2||, \ldots, ||x_d||)^T \in R^d \tag{5.22}$$

where $||.||$ denotes the norm in X. Further, the norm in R^q is defined by

$$||x||_q = \max_{1 \le i \le q} |x_i| \tag{5.23}$$

where $x \in R^q$ is regarded as the column $x = (x_1, x_2, \ldots, x_q)^T$, and the norm in X^d is defined by

$$||x|| = \max_{1 \le i \le d} ||x_i|| \tag{5.24}$$

∎

Definition 5.1.3: Let $B(X^{d_2}, X^{d_1})$ denote the space of bounded linear operators mapping X^{d_2} into X^{d_1}. Further, represent $L \in B(X^{d_2}, X^{d_1})$ as

$$Y = Lx \tag{5.25}$$

or

$$y_i = \sum_j L_{ij} x_j \tag{5.26}$$

where the L_{ij} are bounded linear operators in X. Then the absolute value of L is defined to be

$$||L||_p = \begin{bmatrix} ||L_{11}|| & & ||L_{1d_2}|| \\ & & \\ ||L_{d_1 1}|| & & ||L_{d_1 d_2}|| \end{bmatrix} \tag{5.27}$$

where $||.||$ is also used to denote the operator norm induced by the vector norm in X.

Extensive use will be made of the following definitions and results for the ∎ special case of $X = L_\infty(0,+\infty)$.

Definition 5.1.4: The extended space of $X^d = L_\infty^d(0,+\infty)$ is denoted by X_e^d. Further, the natural projection of $L \in X_e^d$ into $X_{(0,T)}^d = L_\infty^d(0,T)$, regarded as a subspace of X^d, is denoted by $P_T L$.

∎

Lemma 5.1.5: Consider $L \in B(X^{d_2}, X^{d_1})$ of (5.25) and suppose that its elements L_{ij} of (5.26) have the convolution form

$$(L_{ij}x_j)(t) = d_{ij}x_j(t) + \int_0^t H_{ij}(t-t')x_j(t')dt' \tag{5.28}$$

Then $P_T L_{ij}$ has induced norm

$$||P_T L_{ij}|| = |d_{ij}| + \int_0^T |H_{ij}(t')|dt' \tag{5.29}$$

in $L_\infty(0,T)$. ∎

Lemma 5.1.6: Suppose that $L \in B(X^{d_2}, X^{d_1})$ has elements of the form (5.28) and denote the step response matrix of L by $Q(t)$ with elements $Q_{ij}(t)$. Then

$$||P_T L_{ij}|| = N_T(Q_{ij}), \ 1 \le i \le d_1, \ 1 \le j \le d_2, \ \forall \ T > 0 \tag{5.30}$$

and hence

$$||P_T L||_p = \begin{bmatrix} N_T(Q_{11}) & & N_T(Q_{1d_2}) \\ & & \\ N_T(Q_{d_1 1}) & & N_T(Q_{d_1 d_2}) \end{bmatrix}, \ \forall \ T > 0 \tag{5.31}$$

∎

Theorem 5.1.1: Suppose that the elements of $L \in B(X^{d_2}, X^{d_1})$ have the structure of (5.28). Then, $\forall \ T > 0$,

$$||P_T L|| = ||(||P_T L||_p)||$$

$$= \max_{1 \le i \le d_1} \sum_{j=1}^{d_2} N_T(Q_{ij})$$

$$\le ||L|| = ||(||P_\infty L||_p)||$$

$$= \max_{1 \le i \le d_1} \sum_{j=1}^{d_2} N_\infty(Q_{ij}) \tag{5.32}$$

∎

5.2 Stability Tests

This section developes simulation-based stability tests for the differential and discrete non-unit memory linear repetitive processes of examples 2.3.3 and 2.3.7 respectively and their unit memory special cases. These tests are based on suitably well behaved plant step response data which is assumed to be available or can be obtained by appropriate simulation studies. In the unit memory case, however, this data could (in principle) be obtained by appropriate experiments on the actual plant or process.

Consider first, therefore, the differential non-unit memory case with state-space model

$$\dot{X}_{k+1}(t) = AX_{k+1}(t) + BU_{k+1}(t) + \sum_{j=1}^{M} B_{j-1}Y_{k+1-j}(t)$$

$$Y_{k+1}(t) = CX_{k+1}(t) + D_0 U_{k+1}(t) + \sum_{j=1}^{M} D_j Y_{k+1-j}(t)$$

$$X_{k+1}(t) \in R^n, \ Y_{k+1}(t) \in R^m, \ U_{k+1}(t) \in R^\ell$$

$$0 \le t \le \alpha, \ X_{k+1}(0) = 0, \ k \ge 0 \tag{5.33}$$

or, on solving the state equation,

$$Y_{k+1}(t) = C \int_0^t e^{A(t-\tau)} \{ \sum_{j=1}^{M} B_{j-1}Y_{k+1-j}(\tau) + BU_{k+1}(\tau) \} d\tau$$

$$+ D_0 U_{k+1}(t) + \sum_{j=1}^{M} D_j Y_{k+1-j}(t), \ 0 \le t \le \alpha, \ k \ge 0 \tag{5.34}$$

Further, consider the problem in the context of the Banach space $E_\alpha = C_m(0,\alpha)$ of bounded continuous mappings of the interval $0 \le t \le \alpha$ into the vector space of real m-vectors R^m with norm

$$||Y|| = \sup_{0 \le t \le \alpha} ||Y(t)||_m \tag{5.35}$$

where $||.||_m$ is any convenient norm in R^m, e.g. $||P||_m = \max_{1 \le i \le m} |P_i|$. Then, as shown in the original analysis of example 2.3.3, (5.33) is a special case of $S(E_\alpha, W_\alpha, L_\alpha)$ in the product space $E_\alpha^M = E_\alpha \times E_\alpha \times \ldots \times E_\alpha$ (M times) with dynamics described by the companion form recursion relations of (2.23)-(2.24). In particular,

$$L_\alpha = \begin{bmatrix} 0 & I & & 0 \\ & & & \\ 0 & & 0 & I \\ L_\alpha^M & L_\alpha^{M-1} & L_\alpha^2 & L_\alpha^1 \end{bmatrix} \tag{5.36}$$

where (see (2.32) and (2.33) respectively) L_α^j, $1 \le j \le M$, is defined by

$$(L_\alpha^j Y)(t) = C \int_0^t e^{A(t-\tau)} B_{j-1}Y(\tau)d\tau + D_j Y(t), \ 0 \le t \le \alpha \tag{5.37}$$

and the disturbance b_{k+1} by

$$b_{k+1} = C \int_0^t e^{A(t-\tau)} BU_{k+1}(\tau)d\tau + D_0 U_{k+1}(t), \ 0 \le t \le \alpha \tag{5.38}$$

The analysis which follows uses as a starting point the so-called associated conventional linear systems of (5.33) defined by (2.53) as

$$\dot{X}(t) = AX(t) + B_{j-1}Y^{1-j}(t)$$

$$W^j(t) = CX(t) + D_j Y^{1-j}(t)$$

$$X(0) = 0, \ 1 \le j \le M \tag{5.39}$$

and it is assumed here that each member of this set is controllable and observable. (Note also again the interpretation of these systems given immediately after (2.53)). Further, the following assumptions concerning the step response matrix of each of these systems are invoked.

Assumption 5.2.1: Write the jth element of (5.39) in the convolution form $W^j = L^j Y^{1-j}$ where

$$(L^j Y^{1-j})(t) = \int_0^t H^j(t') Y^{1-j}(t-t') dt' + D_j Y^{1-j}(t) \tag{5.40}$$

and $H(t)$ is the $m \times m$ impulse response matrix

$$H^j(t) = C e^{At} B_{j-1} \tag{5.41}$$

Then it is assumed that the step response matrix

$$W^j(t) = \int_0^t H^j(t') dt' + D_j, \quad t \geq 0 \tag{5.42}$$

of this element is available and it is convenient to write this matrix in the form

$$W^j(t) = \begin{bmatrix} W^j_{11}(t) & W^j_{1m}(t) \\ W^j_{m1}(t) & W^j_{mm}(t) \end{bmatrix} \tag{5.43}$$

Here $W^j_{pv}(t)$ denotes the response of the pth output channel to a unit step in the vth input channel.

Assumption 5.2.2: $W^j(t)$ is assumed to be a stable response. Formally, it is required that

$$||W^j(t)||_m \leq \int_0^\infty ||H^j(t')||_m dt' + ||D_j||_m < +\infty \tag{5.44}$$

where $||.||_m = \max_i \Sigma_j |(.)_{ij}|$ is the matrix norm induced by the vector norm $||.||_m = \max_i |(.)_i|$ in R^m.

Note that for (5.44) to hold all eigenvalues of the matrix A must have strictly negative real parts. This is also, see (b) of theorem 3.3.7, a necessary condition for stability along the pass. Further, it is assumed here that $W^j(t)$ is available from appropriate simulation studies on the jth, $1 \leq j \leq M$, element of (5.39). In the unit memory case, however, the corresponding matrix could (in principle) be obtained by appropriate experiments on the actual plant or process.

Suppose now that $E_\alpha = C_m(0,\alpha)$ is replaced by $E_\alpha = L^m_\infty(0,+\infty)$ in the abstract model $S(E_\alpha, W_\alpha, L_\alpha)$ where L_α has the block companion structure of (5.36)-(5.37). Further, define $L \in B(X^N, X^N)$, $X^N = L^N_\infty(0,+\infty)$, $N = mM$, as

$$L \quad = \begin{bmatrix} 0 & I & & 0 \\ 0 & 0 & & I \\ L^M & L^2 & & L^1 \end{bmatrix} \tag{5.45}$$

In which case it follows immediately that the natural projection, see definition 5.1.4, of $L \in X_e^N$ into $X_{(0,\alpha)}^N = L_\infty^N(0,\alpha)$ is just L_α of (5.36), i.e.

$$P_\alpha L = L_\alpha , \quad 0 < \alpha < +\infty \tag{5.46}$$

and

$$P_\infty L = L \tag{5.47}$$

Now apply the result of lemma 5.1.1 to each element in turn of $W^j(t)$, $1 \le j \le M$, to construct the matrix $||P_\infty L^j||_p$ of (5.31) and hence the N×N block companion matrix

$$||L||_p = \begin{bmatrix} 0 & I_m & & 0 \\ & & & \\ 0 & 0 & & I_m \\ ||L^M||_p & ||L^2||_p & & ||L^1||_p \end{bmatrix} \tag{5.48}$$

Then it follows immediately that the following application of the partial ordering of definition 5.1.1 holds

$$||L_\alpha||_p \le ||L||_p , \quad 0 < \alpha < +\infty \tag{5.49}$$

The following is the central result of this section and expresses stability along the pass of example 2.3.3 in terms of the matrix $||L||_p$ of (5.48).

Theorem 5.2.1: Suppose that the matrix $||L||_p$ of (5.48) has been constructed for the differential non-unit memory linear repetitive process of example 2.3.3. Then the extended linear repetitive process $S(E_\alpha, W_\alpha, L_\alpha)_{\alpha \ge \alpha_0}$ generated by the model of this example with $\alpha \ge \alpha_0$ is stable along the pass if

$$r(||L||_p) < 1 \tag{5.50}$$

Proof: In effect, this consists of showing that (5.50) is a sufficient condition for the general stability along the pass result of theorem 3.3.2 to hold, i.e.

$$\text{(a)} \quad r_\infty = \sup_{\alpha \ge \alpha_0} r(L_\alpha) < 1 \tag{5.51}$$

and

$$\text{(b)} \quad M_0 := \sup_{\alpha \ge \alpha_0} \sup_{|z| \ge \lambda} ||(zI - L_\alpha)^{-1}|| < +\infty \tag{5.52}$$

for some real number $\lambda \in (r_\infty, 1)$. Further, in the case of (5.51) note again the partial ordering of (5.49) and apply (v) of lemma 5.1.3 to yield

$$r(L_\alpha) \le r(||L_\alpha||_p) \le r(||L||_p) < 1, \quad 0 < \alpha < +\infty \tag{5.53}$$

Hence (5.50) is clearly a sufficient condition for (5.51) to hold.

To prove (b), first note that if $r(L_\alpha) < |z|$ then $\frac{1}{z}(I - \frac{L_\alpha}{z})^{-1}$ can be represented by the absolutely convergent power series

$$\frac{1}{z}(I - \frac{L_\alpha}{z})^{-1} = \frac{1}{z}(I + \frac{L_\alpha}{z} + (\frac{L_\alpha}{z})^2 + \ldots) \tag{5.54}$$

Further, for $0 < \alpha < +\infty$,

$$||(zI - L_\alpha)^{-1}||_p \leq \frac{1}{|z|} \sum_{j \geq 0} \frac{||L_\alpha^j||_p}{|z|^j} \tag{5.55}$$

$$\leq \frac{1}{|z|} \sum_{j \geq 0} \frac{||L_\alpha||_p^j}{|z|^j}$$

$$\leq \frac{1}{|z|} \sum_{j \geq 0} \frac{||L||_p^j}{|z|^j} \tag{5.56}$$

on application of (i)-(iv) of lemma 5.1.3 to (5.54) and use of the partial ordering (5.49). Now consider the term

$$\hat{M} = \frac{1}{|z|} \sum_{j \geq 0} \frac{||L||_p^j}{|z|^j} \tag{5.57}$$

of (5.56) and note that (5.50) implies the existence of a real number λ in the range $r_\infty \leq r(||L||_p) < \lambda < 1$ and hence

$$\hat{M} \leq \frac{1}{\lambda} \sum_{j \geq 0} \frac{||L||_p^j}{\lambda^j}$$

$$= (\lambda I_N - ||L||_p)^{-1} := M < +\infty \tag{5.58}$$

Combining (5.58) and (5.56) yields

$$||(zI - L_\alpha)^{-1}||_p \leq M, \quad 0 < \alpha < +\infty, \quad \forall \, |z| \geq \lambda \tag{5.59}$$

Taking the norm of (5.59) now leads immediately to the conclusion that

$$\sup_{\alpha \geq \alpha_0} \sup_{|z| \geq \lambda} ||(||(zI - L_\alpha)^{-1}||_p)|| = \sup_{\alpha \geq \alpha_0} \sup_{|z| \geq \lambda} ||(zI - L_\alpha)^{-1}||$$

$$\leq ||M|| < +\infty \tag{5.60}$$

and the proof is complete. ∎

At this stage, note that the initial entries in $W^j(t)$, $1 \leq j \leq M$, of (5.42) or (5.43) are simply the elements of the matrix D_j and hence the entries in

$$||D||_p = ||\lim_{T \to 0+}(P_T L)||_p \tag{5.61}$$

are given by

$$||D||_p = \begin{bmatrix} 0 & I_m & & 0 \\ & & & \\ 0 & 0 & & I_m \\ ||D_M||_p & ||D_2||_p & ||D_1||_p \end{bmatrix} \tag{5.62}$$

Further, note again from (a) of theorem 3.3.7 that (5.33) is asymptotically stable, and hence a necessary condition for stability along the pass holds, if, and only if, the spectral radius of the matrix D is strictly less than unity. Application of the spectral radius inequality $r(D) \leq r(||D||_p)$ (use (v) of lemma 5.1.3) now leads immediately to the following result which, given (5.49) and (v) of lemma 5.1.3, is, in effect, a simple preliminary test for the applicability of theorem 5.2.1 to a given example.

Lemma 5.2.1: The differential non-unit memory linear repetitive process of example 2.3.3 is asymptotically stable if

$$r(||D||_p) < 1 \tag{5.63}$$

where the matrix $||D||_p$ is defined by (5.62). ∎

Given theorem 5.2.1 and lemma 5.2.1 the following steps now, in effect, represent a systematic stability test procedure for example 2.3.3 in the general case.

STEP 1: Obtain the $W^j(t)$, $1 \leq j \leq M$, of (5.42) or (5.43) by appropriate simulation studies on the associated conventional linear systems of (5.39). Experience has shown that the 4th order Runge Kutta method for the numerical integration gives sufficient accuracy in most cases. For comprehensive details of this aspect, see the cited reference which reports progress to date on the development of a comprehensive computer aided analysis/design package for examples 2.3.3 and 2.3.7.

STEP 2: Compute $||D||_p$ of (5.62) either directly (since the entries in the matrices D_j, $1 \leq j \leq M$, are assumed known here) or by use of (5.61) and hence test lemma 5.2.1. Stop if this lemma does not hold since it is, in effect, a preliminary test for the applicability of theorem 5.2.1.

STEP 3: Compute the matrix $||P_\infty L^j||_p \equiv ||L^j||_p$ of (5.31) and hence N × N block companion matrix $||L||_p$ of (5.48), by applying the result of lemma 5.1.1 to each element in turn of $W^j(t)$.

STEP 4: Compute $r(||L||_p)$ and the example under consideration is stable along the pass by theorem 5.2.1 if $r(||L||_p) < 1$.

Suppose, therefore, that $W^j(t)$, $1 \leq j \leq M$, is available and step 2 yields a positive answer. Then, in effect, the above procedure reduces to the computation of $||P_\infty L^j||_p$, followed by the construction of $||L||_p$ and the evaluation of its spectral radius. Hence the computationally feasible stability test arising from theorem 5.2.1 is suitable for software implementation and therefore for inclusion within a CAD package. (See also the cited reference). Note also that a number of special cases exist where it is possible to obtain an explicit formula for $||P_\infty L^j||_p$ with the consequent possibility of obtaining 'synthesis type' results for use in design studies. The following examples illustrate this and other features of the application of theorem 5.2.1 using the above systematic procedure.

<u>Example 5.2.1</u> - The Monotonic Sign Definite Case - Suppose that entry (p,q), $1 \leq p,q \leq m$, in $W^j(t)$, $1 \leq j \leq M$, is monotonic and sign definite. Then it follows immediately that only its steady state value is required to compute the corresponding entry in $||P_\infty L^j||_p$. Further, if all entries in $W^j(t)$ are monotonic and sign definite then

$$||L^j||_p = ||W^j(+\infty)||_p \tag{5.64}$$

which is particularly easy to compute. Alternatively, since the entries in the matrices of its state-space model are assumed known here, write the jth element of (5.39) in transfer-function matrix terms and apply the final-value theorem to yield

$$||W^j(+\infty)||_p = ||G_j(0)||_p = ||-CA^{-1}B_{j-1} + D_j||_p \tag{5.65}$$

A number of practically relevant special cases exist where (5.64) and (5.65) can be used to great effect. For example, consider the SISO unit memory sub-class where the transfer-function of associated conventional linear system has the form

$$G_1(s) = \beta \frac{\prod\limits_{i=1}^{n-1}(s - z_i)}{\prod\limits_{i=1}^{n}(s - \lambda_i)} \tag{5.66}$$

where β is a positive real scalar. Suppose also that the z_i, $1 \leq i \leq n-1$ and the λ_i, $1 \leq i \leq n$, are real, distinct and negative and satisfy the interlacing condition

$$\lambda_1 < z_1 < \lambda_2 < \ldots < z_{n-1} < \lambda_n \tag{5.67}$$

Then it follows immediately on considering the coefficients in the partial fraction expansion of $\dfrac{G_1(s)}{s}$ that $W^1(t)$ is monotonic and sign definite in this case. Hence (since $||L||_p \equiv ||L^1||_p$ in the unit memory case)

$$||L||_p = \beta \left| \frac{\prod\limits_{i=1}^{n-1} z_i}{\prod\limits_{i=1}^{n} \lambda_i} \right| \tag{5.68}$$

and use of theorem 5.2.1 yields stability along the pass if

$$\beta \left| \prod_{i=1}^{n-1} z_i \right| < \left| \prod_{i=1}^{n} \lambda_i \right| \tag{5.69}$$

Example 5.2.2 - The SISO Underdamped Second Order Lag Case - Let the process under consideration be SISO and, for simplicity, unit memory. Suppose also that the associated conventional linear system transfer-function is an underdamped second order lag, i.e.

$$G_1(s) = \frac{\beta \, \omega_n^2}{s^2 + 2\gamma\omega_n s + \omega_n^2} \tag{5.70}$$

where $\gamma \in (0,1)$ is the damping ratio, ω_n is the undamped natural frequency, and β is a positive real scalar. Then in this case $||L||_p$ can be computed from the closed form

$$||L||_p = \beta \frac{(1 + e^{-a})}{1 - e^{-a}} \tag{5.71}$$

where

$$a = \frac{\gamma\pi}{\sqrt{1 - \gamma^2}} \tag{5.72}$$

To prove (5.71), first note that the unit step response in this case is easily shown to be given by

$$W^1(t) = \beta \left(1 + \frac{\bar{\lambda}e^{-\lambda t}}{(\lambda - \bar{\lambda})} + \frac{\lambda e^{-\bar{\lambda}t}}{(\bar{\lambda} - \lambda)} \right) \tag{5.73}$$

where

$$\lambda = -\gamma \, \omega_n - i \, \omega_n \sqrt{1 - \gamma^2} \tag{5.74}$$

and $\bar{\lambda}$ is the complex conjugate. Further, solving the equation $\ddot{W}^1(t) = 0$ yields the time sequence

$$t_k = \frac{k\pi}{\omega_n \sqrt{1 - \gamma^2}} \quad , \quad k = 0,1,2,\ldots \tag{5.75}$$

and hence, using (5.73) and (5.75),

$$W^1(t_k) = \beta \left(1 + (-1)^k 2A \, e^{-\gamma\omega_n t_k} \cos \phi \right) \tag{5.76}$$

where A and ϕ are deduced from the polar decomposition

$$\frac{\bar{\lambda}}{\lambda - \bar{\lambda}} = A \, e^{i\phi} \tag{5.77}$$

Hence $W^1(t_o) = 0$ and

$$W^1(t_k) - W^1(t_{k-1}) = \beta((-1)^k 2A \ e^{-(k-1)a}(e^{-a}+1)\cos \phi), \ k \geq 1 \tag{5.78}$$

which immediately yields

$$||L||_p = \beta((1 + e^{-a})|2A \cos \phi| \sum_{k=1}^{\infty} e^{-(k-1)a}) \tag{5.79}$$

The closed form of (5.71) now follows immediately on summing the infinite geometric series in (5.79). Use of theorem 5.2.1 now yields stability along the pass in this case if

$$\beta(1 + e^{-a}) < 1 - e^{-a} \tag{5.80}$$

Finally, note that if $\gamma \geq 1$ (the critically damped and overdamped cases respectively) then example 5.2.1 applies and $||L||_p = \beta$.

Example 5.2.3 - A Numerical Study - To illustrate one element of the computer aided analysis/design package referred to earlier in this section, consider the unit memory process described by

$$\dot{X}_{k+1}(t) = \begin{bmatrix} -1.6 & -4 \\ 1 & 0 \end{bmatrix} X_{k+1}(t) + \begin{bmatrix} 1 & 0 \\ 0 & 1 \end{bmatrix} U_{k+1}(t) + \begin{bmatrix} 0.33 & 0 \\ 0 & 0.66 \end{bmatrix} Y_k(t)$$

$$Y_{k+1}(t) = \begin{bmatrix} 1 & 1 \\ 0 & 1 \end{bmatrix} X_{k+1}(t)$$

$$0 \leq t \leq \alpha, \ X_{k+1}(0) = 0, \ k \geq 0 \tag{5.81}$$

The associated conventional linear system in this case has transfer-function matrix

$$G_1(s) = \frac{1}{s^2 + 1.6s + 4} \begin{bmatrix} 0.33s + 0.33 & 0.66s - 1.584 \\ 0.33 & 0.66s + 1.056 \end{bmatrix} \tag{5.82}$$

and the elements of $W^1(t)$ are shown in Figure 5.2. Application of lemma 5.1.1 to each of these in turn yields

$$||L||_p = \begin{bmatrix} 0.278 & 0.903 \\ 0.139 & 0.636 \end{bmatrix} \tag{5.83}$$

Further,

$$r(||L||_p) = 0.854 \tag{5.84}$$

and this process is stable along the pass by theorem 5.2.1.

At this stage, restrict attention to the unit memory case. Then for such processes an alternative version of theorem 5.2.1 exists which will find particular use in the next section. To derive this result, first suppose that $||L||_p \equiv ||L^1||_p$ has been constructed and apply theorem 5.1.1 to compute the scalar $||L||$ as

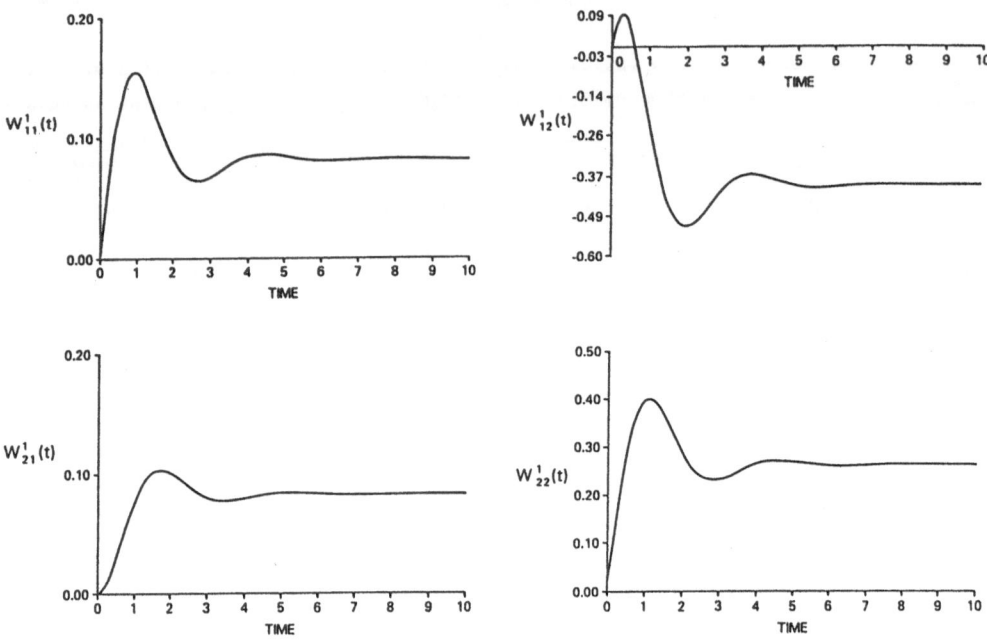

FIGURE 5.2

$$||L|| = ||(||L||_p)|| = \max_{1 \leq i \leq m} \sum_{j=1}^{m} N_\infty(W_{ij}^1(t)) \tag{5.85}$$

Further, combining the spectral radius inequality

$$r(||L||_p) \leq ||(||L||_p)|| = ||L|| \tag{5.86}$$

with (v) of lemma 5.1.3 yields

$$r(L_\alpha) \leq r(||L_\alpha||_p) \leq r(||L||_p) \leq ||L||, \; 0 < \alpha < + \infty \tag{5.87}$$

and it follows immediately that $||L|| < 1$ is a sufficient condition for stability along the pass in this case. A result which is stated formally as follows.

Theorem 5.2.2: Suppose that $||L||$ of (5.85) has been computed for the differential unit memory linear repetitive process of example 2.3.4. Then the extended linear repetitive process $S(E_\alpha, W_\alpha, L_\alpha)_{\alpha \geq \alpha_0}$ generated by the model of this example with $\alpha \geq \alpha_0$ is stable along the pass if

$$||L|| < 1 \tag{5.88}$$

The result of theorem 5.2.2 in its current form cannot be applied to non-unit memory processes since the block companion structure of (5.48) in this case means that $||L||$ is always at least equal to unity. Further, in the unit memory case it follows immediately from (5.87) that theorem 5.2.1 is less convervative than theorem 5.2.2 in the sense that it can be applied to a wider class of examples. In particular, it can be applied to processes where $r(||L||_p) < 1$ but $||L|| \geq 1$, a situation which is highlighted by the following example where β is a real scalar.

$$\dot{X}_{k+1}(t) = \begin{bmatrix} -3 & 0 \\ \beta & -2 \end{bmatrix} X_{k+1}(t) + \begin{bmatrix} 1 & 0 \\ 0 & 1 \end{bmatrix} U_{k+1}(t) + \begin{bmatrix} 1 & 0 \\ 0 & 1 \end{bmatrix} Y_k(t)$$

$$Y_{k+1}(t) = \begin{bmatrix} 1 & 0 \\ 0 & 1 \end{bmatrix} X_{k+1}(t)$$

$$0 \leq t \leq \alpha, \; X_{k+1}(0) = 0, \; k \geq 0 \tag{5.89}$$

It is easily shown using, for example, the first (eigenvalue based) systematic test procedure developed in section 4.2 that (5.89) is stable along the pass for all possible choices of β. Alternatively, suppose that theorem 5.2.1 or 5.2.2 is to be used. Then the associated conventional linear system of (5.89) has transfer-function matrix

$$G_1(s) = \begin{bmatrix} 1/s+3 & 0 \\ \beta/(s+3)(s+2) & 1/s+2 \end{bmatrix} \tag{5.90}$$

Further, inspection of the non-zero elements in $G_1(s)$ immediately indicates that they all have monotonic sign definite step responses and hence, by (5.64) and (5.65) of example 5.2.1,

$$||L||_p = \begin{bmatrix} 1/3 & 0 \\ \beta/6 & 1/2 \end{bmatrix} \tag{5.91}$$

Hence theorem 5.2.1 holds since $r(||L||_p) = \dfrac{1}{2}$ and theorem 5.2.2 holds if

$$||L|| = ||(||L||_p)|| = \frac{1}{2} + \beta/6 < 1 \tag{5.92}$$

In this case, therefore, theorem 5.2.1 holds for all possible choices but theorem 5.2.2 produces an inconclusive result for all $\beta \geq 3$.

Note: The constraint on applying theorem 5.2.2 to non-unit memory processes can (in principle) be removed by use of appropriate similarity transformations. This particular aspect requires much further development work, however, and is left here as a topic for possible future research.

The tests based on theorem 5.2.1 or 5.2.2 are sufficient, but not necessary, and examples are easily generated where they produce an inconclusive result. Further, the stability along the pass characteristics of such examples can only be determined by using necessary and sufficient tests such as those developed in chapter 4. Consequently these tests are not as generally applicable as their counterparts of chapter 4. Suppose, however, that theorem 5.2.1, or theorem 5.2.2 in the unit memory case, holds for the particular example under consideration. Then it will be shown in the next section that these tests produce, at no extra cost, computable information concerning the rate of approach to the limit profile, together with bounds on the performance along any pass, in one case of major practical interest. This information is unique to these tests and the next chapter considers its use in the formulation of controller design algorithms.

To conclude this section, the analysis below develops a refinement of the above tests based, essentially, on filtering the elements in the step response matrices of the associated conventional linear systems. Following this, the extension of all the results developed in this section to the discrete non-unit memory process of example 2.3.7, or its unit memory version of example 2.3.8, is noted. In the case of the former topic, it is shown how the tests developed to date in this section can be refined by filtering $W^j(t)$, $1 \leq j \leq M$, of (5.39) with a filter defined in frequency domain, or transfer-function, terms. Its effective basis is the following result.

Lemma 5.2.2: Suppose that L is a bounded linear convolution operator mapping $L_\infty(0,+\infty)$ into itself with transfer-function $L(s)$. Then

$$|L(s)| \leq N_\infty(Y_L), \quad \forall \, \text{Re}\{s\} \geq 0 \tag{5.93}$$

where $Y_L(t)$ denotes the unit step response of L.

<u>Proof</u>: Write Y = LU as

$$Y(t) = \int_0^t H(t')U(t-t')dt' \tag{5.94}$$

and note that

$$Y_L(t) = \int_0^t H(t')dt' \tag{5.95}$$

Further,

$$L(s) = \int_0^\infty e^{-st}H(t)dt \tag{5.96}$$

and hence

$$|L(s)| \le \int_0^\infty |e^{-st}||H(t)|dt \le \int_0^\infty |H(t)|dt, \quad \forall \text{ Re}\{s\} \ge 0 \tag{5.97}$$

The result now follows immediately on using the total variation result of lemma 5.1.1. ∎

Consider now the SISO unit memory case and suppose that the step response of the associated conventional linear system satisfies assumption 5.2.2 or, equivalently, all eigenvalues of the matrix A have strictly negative real parts. In which case it follows immediately from lemma 5.2.2 that

$$|G_1(i\omega)| \le ||G_1|| \le N_\infty(W^1), \quad \forall \text{ real } \omega \tag{5.98}$$

where

$$||G_1|| = \sup_\omega |G_1(i\omega)| \tag{5.99}$$

Further, it follows immediately from corollary 3.3.10 that this special case is stable along the pass if, and only if,

$$||G_1|| < 1 \tag{5.100}$$

Suppose also that $W^1(t)$ is monotonic and sign definite. Then the following is a useful preliminary result which strengthens theorem 5.2.1 (or theorem 5.2.2) to a necessary and sufficient condition in this special case.

<u>Lemma 5.2.3</u>: Consider the case when the unit memory process of example 2.3.4 is SISO. Suppose also that the step response of the associated conventional linear system is monotonic and sign definite and $N_\infty(W^1) = |W^1(+\infty)|$ has been computed. Then the extended linear repetitive process $S(E_\alpha, W_\alpha, L_\alpha)_{\alpha \ge \alpha_0}$ generated by the model in this case with $\alpha \ge \alpha_0$ is stable along the pass if, and only if,

$$N_\infty(W^1) < 1 \tag{5.101}$$

<u>Proof</u>: Sufficiency follows immediately from (5.98)-(5.100). To show necessity, it is required to prove that

$$||G_1|| = N_\infty(W^1) = |W^1(+\infty)| \tag{5.102}$$

where $N_\infty(W^1) = |W^1(+\infty)|$ follows from example 5.2.1. Further, $||G_1|| \leq N_\infty(W^1)$ by (5.98) and (5.102) follows immediately since $G_1(0) = W^1(+\infty)$ by definition. ∎

Note: Using lemma 5.2.3, (5.69) is necessary and sufficient for stability along the pass of the sub-class of example 2.3.4 whose associated conventional linear system is defined by (5.66).

Continuing with the SISO unit memory case, return to (5.98)-(5.100) and discard the monotonic sign definite assumption. Then $N_\infty(W^1) < 1$ is an upper bound on the frequency dependent necessary and sufficient condition $||G_1|| < 1$ for stability along the pass. Further, it is clearly the best frequency independent upper bound on this condition. Hence it is to be expected that there is an infinite number of frequency dependent upper bounds on $||G_1||$ which can be used to refine the sufficient condition $N_\infty(W^1) < 1$. The following result is used below to characterise a class of bounds which can be obtained by filtering operations on $W^1(t)$. Note that these bounds do not require detailed knowledge of $G_1(s)$. (Recall from the discussion immediately after assumption 5.2.2 that $W^1(t)$ can (in principle) be obtained by appropriate experiments on the actual plant or process.)

Lemma 5.2.4: Let L be a bounded linear convolution operator mapping $L_\infty(0,+\infty)$ into itself with transfer-function $L(s)$. Suppose also that $Y_L(t)$ denotes the unit step response of L and let F_β be a filter with the properties that:

(a) $Y_L^\beta := F_\beta Y_L \in L_\infty(0,+\infty)$; and

(b) $F_\beta^{-1}(s)$ is bounded and analytic in the open right-half plane.

Then

$$|L(s)| \leq \Delta_\beta(s) := |F_\beta^{-1}(s)| N_\infty(Y_L^\beta), \quad \forall \text{ Re}\{s\} \geq 0 \tag{5.103}$$

Proof: Write $L = F_\beta^{-1}(F_\beta L)$ and apply lemma 5.2.2 to $F_\beta L$. ∎

Applying lemma 5.2.4 to the SISO version of example 2.3.4 now yields stability along the pass if

$$||G_1|| \leq \Delta < 1 \tag{5.104}$$

where

$$\Delta := \sup_\omega |F_\beta^{-1}(i\omega)| N_\infty(W_\beta^1) \tag{5.105}$$

and $N_\infty(W_\beta^1)$ is the total variation of $F_\beta W^1$. Further, the choice of $F_\beta = 1$ reduces this new condition to that of theorem 5.2.1 (or theorem 5.2.2), i.e.

$$||G_1|| \leq N_\infty(W^1) < 1 \tag{5.106}$$

In general, however, F_β yields a frequency dependent bound capable of producing more refined results which approach the necessary and sufficient condition of (5.100) for this case. For example, if $F_\beta(s) = G_1^{-1}(s)$ then $W_\beta^1(t) \equiv 1$ and hence (5.105) reduces to (5.106) in this particular case. In practice, however, this choice of $F_\beta(s)$ is not available but it is intuitively obvious that other choices, with a simpler structure, can be used to produce easily computed intermediate estimates for the upper bound on $||G_1||$ imposed by $N_\infty(W^1)$.

The parameter Δ in (5.104) can be replaced by one obtained from the use of a collection, or set, of filters. In particular, suppose that $\{F_i\}_{1 \le i \le \beta}$ denotes a set of filters satisfying the conditions of lemma 5.2.4. Then it follows immediately that the SISO version of example 2.3.4 is stable along the pass if

$$||G_1|| \le \Delta_A < 1 \tag{5.107}$$

where

$$\Delta_A := \inf_{1 \le i \le \beta} \Delta_i \tag{5.108}$$

and Δ_i is computed as per Δ of (5.105) with F_β replaced by F_i and $N_\infty(W_\beta^1)$ by $N_\infty(F_i W^1)$. Finally, combining (5.106) and (5.107) leads to the following result which summarises the potential refinement of theorem 5.2.1 (or theorem 5.2.2) possible for the SISO unit memory case by appropriate use of the filtering operations detailed here. Its effective operating range is for processes where $\Delta_A < 1$ but $N_\infty(W^1) \ge 1$.

Lemma 5.2.5: The extended linear repetitive process $S(E_\alpha, W_\alpha, L_\alpha)_{\alpha \ge \alpha_0}$ generated by the SISO version of the model of example 2.3.4 with $\alpha \ge \alpha_0$ is stable along the pass if

$$\min[\Delta_A, N_\infty(W^1)] < 1 \tag{5.109}$$

Profitable use of these filtering operations requires, of course, the choice of an appropriate filter set $\{F_i\}_{1 \le i \le \beta}$. Further, the development of rules for choosing this set is still an open question and is not considered further here. Instead, the following example is presented to illustrate the benefits arising from a suitable choice.

The example considered is the case where

$$G_1(s) = \frac{(10 + 0.5s)s + 0.5}{(1 + 4s)(1 + 5s)} \tag{5.110}$$

whose unit step response $W^1(t)$ is omitted here. Application of available software yields $N_\infty(W^1) = 1.64$ and hence theorem 5.2.1 (or theorem 5.2.2) produces an inconclusive result. Consider, therefore, the use of the filter

$$F_\beta(s) = \frac{0.1(1 + 4.5s)^2}{s(1 + \gamma s)} \tag{5.111}$$

where γ is a positive real scalar. Figure 5.3 then shows the graphs of $\Delta_A(s)$ for the particular choices of $\gamma = 0.1$ and 1.0 which have been chosen here for

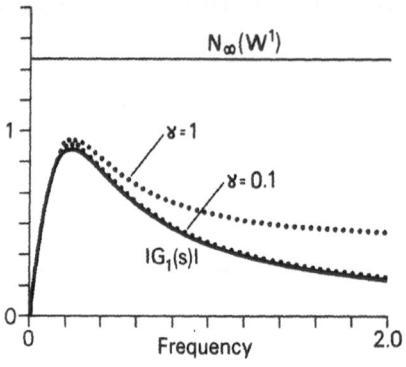

FIGURE 5.3

illustrative purposes only. It now follows immediately that $\Delta_A < 1$ for both of these cases and hence stability along the pass by lemma 5.2.5.

To generalise this filtering analysis, consider first the multivariable version of the unit memory process of example 2.3.4. Suppose also that the step response matrix of the associated conventional linear system satisfies assumption 5.2.2 or, equivalently, all eigenvalues of the matrix A have strictly negative real parts. In which case applying lemma 5.2.2 to each element in turn of $W^1(t)$ yields

$$||G_1(s)||_p \leq ||L||_p , \quad \forall \; Re\{s\} \geq 0 \qquad (5.112)$$

Use of (v) of lemma 5.1.3 now immediately yields

$$r(G_1(i\omega)) \leq \tilde{r} \leq r(||L||_p) , \quad \forall \; real \; \omega \qquad (5.113)$$

where

$$\tilde{r} = \sup_{\omega} r(||G_1(i\omega)||_p) \qquad (5.114)$$

Further, it is easily concluded from corollary 3.3.10 that (note again the assumption on the eigenvalues of the matrix A) any real number in the range $\tilde{r} \leq x \leq r(||L||_p)$ satisfying $x < 1$ is a sufficient condition for stability along the pass.

It is easily shown that the result of theorem 5.2.1, i.e. $r(||L||_p) < 1$ is the best frequency independent upper bound on the necessary and sufficient conditions of corollary (3.3.10) for stability along the pass. Further, it is to be expected that there is an infinite number of frequency dependent upper bounds on these conditions which can be used to refine the result of theorem 5.2.1. The following analysis characterises a class of bounds which can be obtained by filtering operations on $W^1(t)$ which are, in effect, just element by element applications of their SISO counterparts derived earlier. Note also that, as in the SISO case, these bounds do not require detailed knowledge of the elements of $G_1(s)$.

Consider, therefore, element (r,q), $1 \leq r \leq m$, $1 \leq q \leq m$, of $W^1(t)$, written $W^1_{r,q}(t)$, and let $\{F^{r,q}_i\}_{1 \leq i \leq \beta}$ denote the corresponding set of filters. Suppose also that the members of this set satisfy the conditions of lemma 5.2.4 in the sense that

(a) $\hat{W}^1_{r,q} := F^{r,q}_i W^1_{r,q} \in L_\infty(0, +\infty)$; and

(b) $(F^{r,q}_i(s))^{-1}$ is bounded and analytic in the open right-half plane.

Further, construct the $m \times m$ matrix $||\hat{\Delta}||_p$ with element (r,q), written $\hat{\Delta}_{r,q}$, given by

$$\hat{\Delta}_{r,q} = \inf_{1 \leq i \leq \beta} \Delta_i \qquad (5.115)$$

where

$$\Delta_i = \sup_{\omega} |(F^{r,q}_i(i\omega))^{-1}| N_\infty(\hat{W}^1_{r,q}) \qquad (5.116)$$

Then it follows immediately that example 2.3.4 is stable along the pass if

$$r(||\tilde{\Delta}||_p) < 1 \tag{5.117}$$

Finally, combining (5.117) with the result of theorem 5.2.1 leads to the following theorem which summarises the potential refinement possible by appropriate use of the filtering operations detailed above.

<u>Theorem 5.2.3:</u> The extended linear repetitive process $S(E_\alpha, W_\alpha, L_\alpha)_{\alpha \geq \alpha_0}$ generated by the model of example 2.3.4 with $\alpha \geq \alpha_0$ is stable along the pass if

$$\min[r(||\tilde{\Delta}||_p), \ r(||L||_p)] < 1 \tag{5.118}$$
■

Alternatively, compute $||\tilde{\Delta}|| = ||(||\Delta||_p)||$. In which case, the following result summarises the potential refinement possible by use of the (less generally applicable) result of theorem 5.2.2.

<u>Theorem 5.2.4:</u> The extended linear repetitive process $S(E_\alpha, W_\alpha, L_\alpha)_{\alpha \geq \alpha_0}$ generated by the model of example 2.3.4 with $\alpha \geq \alpha_0$ is stable along the pass if

$$\min[||\tilde{\Delta}||, \ ||L||] < 1 \tag{5.119}$$
■

Finally, note that the effective operating range of theorem 5.2.3 is for processes where $r(||\tilde{\Delta}||_p) < 1$ but $r(||L||_p) \geq 1$ and that of theorem 5.2.4 is for processes where $||\tilde{\Delta}|| < 1$ but $||L|| \geq 1$. Further, note also that profitable use of these filtering operations requires the development of rules for choosing appropriate filter sets. Here, however, the development of such rules is, as in the SISO case, left as a future research area.

The filtering operations detailed above extend in a straightforward manner to the non-unit memory process of example 2.3.3. In particular, suppose that a filter set, specified as in the unit memory case, is applied to each element of the step response matrix $W^j(t)$, $1 \leq j \leq M$, of the jth associated conventional linear system and denote the resulting matrix by $||\tilde{\Delta}_j||_p$. Further, construct the following N×N block companion matrix from the $||\tilde{\Delta}_j||_p$, $1 \leq j \leq M$

$$||\tilde{\Delta}||_p = \begin{bmatrix} 0 & I_m & & & 0 \\ & & & & \\ & & & & \\ 0 & & 0 & & I_m \\ ||\tilde{\Delta}_M||_p & & ||\tilde{\Delta}_2||_p & ||\tilde{\Delta}_1||_p \end{bmatrix} \tag{5.120}$$

Then the following result is the formal statement of theorem 5.2.3 for the non-unit memory case.

<u>Theorem 5.2.5:</u> The extended linear repetitive process $S(E_\alpha, W_\alpha, L_\alpha)_{\alpha \geq \alpha_0}$ generated by the model of example 2.3.3 with $\alpha \geq \alpha_0$ is stable along the pass if

$$\min[r(||\tilde{\Delta}||_p), \ r(||L||_p)] < 1 \tag{5.121}$$

where $||\tilde{\Delta}||_p$ is defined by (5.120) and $||L||_p$ by (5.48). ∎

As the final item in this section, consider the discrete non-unit memory linear repetitive process of example 2.3.7 with state-space model

$$X_{k+1}(P+1) = \Phi \ X_{k+1}(P) + \Delta \ U_{k+1}(P) + \sum_{j=1}^{M} \Delta_{j-1} Y_{k+1-j}(P)$$

$$Y_{k+1}(P) = C \ X_{k+1}(P) + D_o U_{k+1}(P) + \sum_{j=1}^{M} D_j Y_{k+1-j}(P)$$

$$X_{k+1}(P) \in R^n, \ Y_{k+1}(P) \in R^m, \ U_{k+1}(P) \in R^\ell$$

$$0 \leq P \leq \alpha, \ X_{k+1}(0) = 0, \ k \geq 0 \tag{5.122}$$

Then, as shown in the analysis of example 2.3.7, (5.122) is a special case of $S(E_\alpha, W_\alpha, L_\alpha)$ in the product space $E_\alpha^M = E_\alpha \times E_\alpha \times \ldots \times E_\alpha$ (M times) with dynamics described by the companion form based recursion relations of (2.23)-(2.24). In particular, E_α is defined by (2.46)-(2.47), L_α by (2.24)(or (5.36)), L_α^j, $1 \leq j \leq M$, by

$$(L_\alpha^j Y)(P) = \sum_{r=0}^{P-1} C\Phi^{P-1-r}\Delta_{j-1}Y(r) + D_j Y(P), \quad 0 \leq P \leq \alpha \tag{5.123}$$

and the disturbance b_{k+1} by

$$b_{k+1} = \sum_{r=0}^{P-1} C\Phi^{P-1-r}\Delta U_{k+1}(r) + D_o U_{k+1}(P), \quad 0 \leq P \leq \alpha \tag{5.124}$$

Further, the associated conventional linear systems of (5.122) are defined by

$$X(P+1) = \Phi \ X(P) + \Delta_{j-1}Y^{1-j}(P)$$

$$W^j(P) = C \ X(P) + D_j Y^{1-j}(P)$$

$$X(0) = 0, \ 1 \leq j \leq M \tag{5.125}$$

Suppose now that each member of the set (5.125) is controllable and observable. (Note again the interpretation of these systems given in section 2.4). Further, introduce the assumptions detailed below concerning the step response matrices of the associated conventional linear systems. Then it follows immediately that all of the results developed in this section extend in a natural manner to (5.122), or its unit memory special case, with all required total variation computations undertaken using lemma 5.1.2. Hence the details are omitted.

Assumption 5.2.3: Write the jth element of (5.125) in the convolution form $W^j = L^j Y^{1-j}$ where

$$(L^j Y^{1-j})(P) = \sum_{r=1}^{P} H^j(r)Y^{1-j}(P-r) + D_j Y(P) \tag{5.126}$$

and $H^j(r)$ is defined by

$$H^j(r) = C\Phi^{r-1}\Delta_{j-1} \tag{5.127}$$

Then it is assumed that the step response matrix

$$W^j(P) = \sum_{r=1}^{P} H^j(r) + D_j, \quad P \geq 0 \qquad (5.128)$$

of this element is available and it is convenient to write this matrix in the form

$$W^j(P) = \begin{bmatrix} W_{11}^j(P) & W_{1m}^j(P) \\ & \\ W_{m1}^j(P) & W_{mm}^j(P) \end{bmatrix} \qquad (5.129)$$

Here $W_{rq}^j(P)$ denotes the response of the rth output channel to a unit step in the qth input channel.

Assumption 5.2.4: $W^j(P)$ is assumed to be a stable response. Formally, it is required that

$$||W^j(P)||_m \leq \sum_{r=1}^{\infty} ||H^j(r)||_m + ||D_j||_m < + \infty \qquad (5.130)$$

where $||.||_m$ is defined as in assumption 5.2.2.

5.3 Performance Bounds

The basic underlying theme of this section is the use of the simulation-based tests of the previous section as a basis for the development of physically meaningful computable performance bounds for examples 2.3.3. and 2.3.7, or their unit memory special cases. In particular, it is shown that these tests produce, at no extra cost, computable information concerning the rate of approach to the limit profile, together with bounds on the performance along any pass, in one special case of major practical interest. This information is unique to these tests and the next chapter considers its use in the formulation of controller design algorithms.

To motivate the results developed in this section, consider the differential process of example 2.3.3 under asymptotic stability and hence, see theorem 3.1.4 and the discussion immediately following, the corresponding limit profile is described by the state-space model

$$\dot{X}_\infty(t) = (A + \hat{B}(I_m - \hat{D})^{-1}C)X_\infty(t) + (B + \hat{B}(I_m - \hat{D})^{-1}D_0)U_\infty(t)$$
$$Y_\infty(t) = (I_m - \hat{D})^{-1}C \, X_\infty(t) + (I_m - \hat{D})^{-1}D_0U_\infty(t) \qquad (5.131)$$

where

$$\hat{B} = \sum_{j=1}^{M} B_{j-1}, \quad \hat{D} = \sum_{j=1}^{M} D_j \qquad (5.132)$$

Hence, in effect, the repetitive dynamics of the process of example 2.3.3 under asymptotic stability can, after a 'sufficiently large' number of passes, be described by a conventional linear systems state-space model. Further, other work, see the cited reference and the next chapter for full details, has considered how this fact can be exploited in terms of the development of physically meaningful control policies and attendant controller design algorithms. In particular, consider the often encountered physical situation where the control sequence applied is constant from pass to pass, i.e. $U_{k+1} = U_\infty$, $k \geq 0$, and hence $b_{k+1} = b_\infty$ in (5.38) of $S(E_\alpha, W_\alpha, L_\alpha)$. Then a number of control policies for this case have been formulated where computable information concerning the following aspects is an essential item:

(i) the rate of approach of the output sequence $\{Y_k\}_{k\geq1}$ to the limit profile Y_∞; and

(ii) bounds for the error $Y_k - Y_\infty$ on any pass $k \geq 0$.

One common choice for U_∞ is

$$U_\infty(s) = \frac{\tilde{e}}{s} \tag{5.133}$$

where at least one element of the column vector $\tilde{e} \in R^\ell$ is unity and the rest are zero. This corresponds to the case where a unit step is applied in one or more input channels at $t = 0$ on each pass. Further, the following theorem is the basic underlying result of this section for the case of example 2.3.3 with a control sequence which is constant from pass to pass.

<u>Theorem 5.3.1</u>: Suppose that the extended linear repetitive process $S(E_\alpha, W_\alpha, L_\alpha)_{\alpha \geq \alpha_0}$ generated by the model of example 2.3.3 with $\alpha \geq \alpha_0$ is stable along the pass and

$$r(||L||_p) < 1 \tag{5.134}$$

where $||L||_p$ is defined by (5.48). Further, let the control sequence applied be constant from pass to pass, i.e. $U_{k+1} = U_\infty$, $k \geq 0$, and hence $b_{k+1} = b_\infty$ in (5.38). Then, for $\alpha \in (0, +\infty)$, there exists an N×N non-negative matrix W and a real scalar $\gamma \in (r(||L||_p), 1)$ such that the error $Y_k - Y_\infty$, $k \geq 0$, satisfies

$$||Y_k - Y_\infty||_p \leq W \gamma^k \{||Y_0||_p + (I_N - ||L||_p)^{-1}||b_\infty||_p\} \tag{5.135}$$

<u>Proof:</u> Since $b_{k+1} = b_\infty$, $k \geq 0$, the solution of the equation describing the dynamics of $S(E_\alpha, W_\alpha, L_\alpha)$, interpreted in the 'companion form' of (2.23)-(2.24), can be written as

$$Y_k = L_\alpha^k Y_0 + \sum_{j=1}^{k} L_\alpha^{j-1} b_\infty \tag{5.136}$$

Further, it is easily shown that the corresponding limit profile can be expressed as

$$Y_\infty = \sum_{j=1}^{\infty} L_\alpha^{j-1} b_\infty \tag{5.137}$$

Hence the error $Y_k - Y_\infty$ can be expressed as

$$Y_k - Y_\infty = L_\alpha^k Y_0 - \sum_{j=k+1}^{\infty} L_\alpha^{j-1} b_\infty \tag{5.138}$$

and therefore

$$||Y_k - Y_\infty||_p \le ||L^k||_p \{||Y_0||_p + \sum_{j=k+1}^{\infty} ||L||_p^{j-1-k} ||b_\infty||_p\} \qquad (5.139)$$

on application of (iii)-(iv) of lemma 5.1.3 and use of the partial ordering relation (5.49).

To proceed, first note that since (5.134) holds then $(I_N - ||L||_p)^{-1}$ exists, is non-negative by lemma 5.1.4, and it is easily shown that

$$(I_N - ||L||_p)^{-1} = \sum_{j=k+1}^{\infty} ||L||_p^{j-1-k} \qquad (5.140)$$

Hence it remains to prove that there exists a non-negative matrix $W \ge 0$ and a real scalar $\gamma \in (r(||L||_p),1)$ such that

$$||L^k||_p \le W \gamma^k, \quad k \ge 0 \qquad (5.141)$$

This follows on noting that $r(L) \le r(||L||_p) < 1$ by (v) of lemma 5.1.3 and therefore it is possible to choose real numbers $\tilde{W} > 0$ and $\gamma \in (r(||L||_p),1)$ such that

$$||L^k|| \le \tilde{W} \gamma^k, \quad k \ge 0 \qquad (5.142)$$

Further, it is clear that the partial ordering $||L^k||_p \le Q$ holds where Q is the N×N matrix with each element equal to $||L^k||$. The result of (5.135) now follows immediately on using (5.140) and defining W as the N×N non-negative matrix with each element equal to $\tilde{W}$. ∎

Suppose, therefore, that the real scalar γ is chosen as any number in the known range $r(||L||_p) < \gamma < 1$. Consider also the output sequence $\{Y_k\}_{k \ge 1}$ in terms of its convergence to the limit profile. Then the computable information available from theorem 5.3.1 is the fact that this sequence approaches Y_∞ at a geometric rate governed by γ. Further, a number of refinements of this result can (in principle) be obtained by use of appropriate filtering operations similar to those detailed in the previous section. This particular aspect requires much further development work and is left here as a possible future research area.

Consider now the unit memory case and suppose that theorem 5.2.2 holds. Then the following result provides an alternative to theorem 5.3.1 for this case.

Theorem 5.3.2: Suppose that the extended linear repetitive process $S(E_\alpha, W_\alpha, L_\alpha)_{\alpha \ge \alpha_0}$ generated by the model of example 2.3.4 with $\alpha \ge \alpha_0$ is stable along the pass and

$$||L|| = ||(||L||_p)|| < 1 \qquad (5.143)$$

Further, let the control sequence applied be constant from pass to pass, i.e. $U_{k+1} = U_\infty$, $k \ge 0$, and hence $b_{k+1} = b_\infty$ in (5.38). Then, for $\alpha \in (0, +\infty)$, the error $Y_k - Y_\infty$, $k \ge 0$, satisfies

$$||Y_k - Y_\infty|| \le ||L||^k \{||Y_0|| + \frac{||b_\infty||}{1 - ||L||}\} \qquad (5.144)$$

Proof: Follows immediately on taking the norm of (5.139). ∎

Suppose, therefore, that theorem 5.3.2 holds and consider the output sequence $\{Y_k\}_{k \geq 1}$ in terms of its convergence to the limit profile. Then the computable information available from theorem 5.3.2 is the fact that this sequence approaches Y_∞ at a geometric rate governed by $||L||$. Further, in common with theorem 5.3.1, a number of refinements of this result can (in principle) be obtained by the use of appropriate filtering operations similar to those detailed in the previous section. This particular aspect requires much further development work and is again left as a possible future research area.

Several potential refinements of theorem 5.3.2 exist which do not require filtering operations. Here, however, only the one based on the fact that $||L^k|| \leq ||L||^k$ is considered and the others are detailed in the cited reference. In this case, if $||L^k|| < ||L||^k < 1$ then $||L^k||$ provides an improved estimate of the rate of convergence of $\{Y_k\}_{k \geq 1}$ to Y_∞. Further, $||L^k||$ is easily computed using the following systematic procedure which is stated here for the SISO case with an obvious generalisation to the multivariable version.

STEP 1: Perform k simulations on $Y^i = LU^i$ where U^1 is a unit step applied at t = 0 and $U^i = Y^{i-1}$, $2 \leq i \leq k$.

STEP 2: Apply lemma 5.1.1 to compute $||L^k||$ as $||L^k|| = N_\infty(Y^k)$.

To illustrate the potential for using theorem 5.3.2 in controller design, with an obvious extension to theorem 5.3.1, return to SISO case defined by (5.66) where, using (5.68),

$$||L|| = \beta \left| \frac{\prod_{i=1}^{n-1} z_i}{\prod_{i=1}^{n} \lambda_i} \right| \tag{5.145}$$

Suppose also that a so-called current pass state feedback control law has been designed, see the cited reference for the necessary theoretical background, which leaves the z_i, $1 \leq i \leq n - 1$, invariant and moves the λ_i, $1 \leq i \leq n$, to locations γ_i which are real, distinct and negative and satisfy the interlacing condition

$$\gamma_1 < z_1 < \gamma_2 < \ldots\ldots < z_{n-1} < \gamma_n \tag{5.146}$$

Further, write the closed-loop associated conventional linear system in convolution form as $W^1 = L_c Y^0$. Then

$$||L_c|| = \beta \left| \frac{\prod_{i=1}^{n-1} z_i}{\prod_{i=1}^{n} \gamma_i} \right| \tag{5.147}$$

and note that $||L_c|| \to 0$ as $\gamma_1 \to -\infty$. Equivalently, stability along the pass coupled with an arbitrary fast rate of approach to the limit profile results from an appropriate choice of γ_i, $1 \leq i \leq n$, satisfying (5.146) (i.e. by placing γ_1 'far enough' to the left of the origin on the real line).

In more general terms, suppose that the number $0 < b < 1$ is available as a measure of the required rate of convergence of $\{Y_k\}_{k \geq 1}$ to Y_∞. Then clearly this requirement can be built into the design procedure as the constraint that $||L_c|| \leq b$. This general area is considered again in the next chapter where a number of control policies and attendant controller algorithms are developed.

Consider now the problem of obtaining bounds for the error $Y_k - Y_\infty$, $k \geq 0$, from theorem 5.3.2. In particular, suppose, for simplicity, that the initial profile, Y_0, is zero and note again the definition of $||.||$. Then it follows immediately that

$$|Y_k^i(t) - Y_\infty^i(t)| \leq ||Y_k - Y_\infty|| \leq \frac{||L||^k ||b_\infty||}{1 - ||L||} , \quad t \geq 0 \tag{5.148}$$

where the notation $Y_k^i(t)$ and $Y_\infty^i(t)$ denotes the ith, $1 \leq i \leq m$, output channel of $Y_k(t)$ and $Y_\infty(t)$ respectively.

Suppose, therefore, that $||b_\infty||$ is obtained as detailed below. Then by (5.148) $Y_k^i(t)$, $t \geq 0$, lies in the 'band' defined by

$$Y_\infty^i(t) - \gamma_k \leq Y_k^i(t) \leq Y_\infty^i(t) + \gamma_k \tag{5.149}$$

where

$$\gamma_k := \frac{||L||^k ||b_\infty||}{1 - ||L||} \tag{5.150}$$

This band has the graphical interpretation shown in Figure 5.4 and it is obviously suitable for inclusion in a CAD environment. Further, its width decreases from pass to pass at a geometric rate governed by $||L||$.

To compute $||b_\infty||$, first note that b_∞ is the response of the derived conventional linear system $L_D(A,B,C,D_0)$ to U_∞, i.e. in state-space terms

$$\dot{X}(t) = AX(t) + BU_\infty(t)$$

$$b_\infty(t) = CX(t) + D_0U_\infty(t)$$

$$X(0) = 0 \tag{5.151}$$

Further, (5.151) is stable since all eigenvalues of the matrix A have strictly negative real parts. Suppose also that $b_\infty(t)$ is available from simulation studies on (5.151) with $U_\infty(t)$. Then it follows immediately that

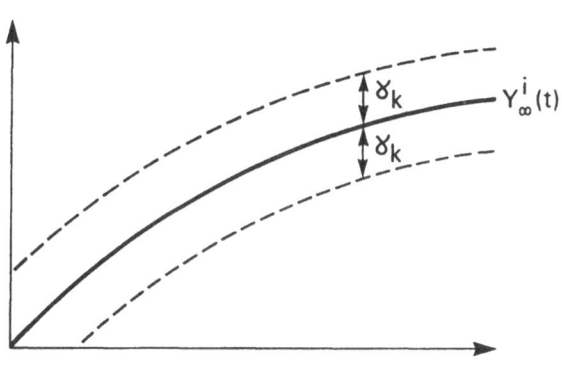

FIGURE 5.4

$$||b_\infty|| = \max_{1 \le i \le m} \sup_{t \ge 0} |b_\infty^i(t)| \qquad (5.152)$$

where $b_\infty^i(t)$ denotes the ith element of $b_\infty(t)$.

In common with the conventional linear systems case, it is to be expected that the systems response will be judged in terms of 'benchmark' choices of $U_\infty(t)$. Consider, therefore, the ith, $i \le i \le m$, channel of $Y_k(t)$. Then one obvious choice, again by analogy with the conventional linear systems case, is

$$U_\infty(s) = \frac{e_i}{s} \qquad (5.153)$$

where e_i is the $\ell \times 1$ column vector consisting of zeros everywhere except the ith position which consists of a unit element. Equivalently, a unit step is applied at $t = 0$ on each pass in this channel with all others identically zero.

An alternative to (5.152) exists in terms of an upper bound on $||b_\infty||$ which, although more conservative in general, may prove computationally more attractive in certain cases. To develop this, first write (5.151) in the convolution form $b_\infty = L_b U_\infty$, where L_b is defined by (5.38) with U_{k+1} replaced by U_∞. Suppose also that the step response matrix, denoted $W_b(t)$, is available from appropriate simulation studies on (5.151) and note that this matrix satisfies assumption 5.2.2. Hence $||L_b||$ can be computed from $W_b(t)$ as per the computation of $||L||$ from $W^1(t)$ in section 5.2. Further,

$$||b_\infty|| \le ||L_b|| \ ||U_\infty|| \qquad (5.154)$$

which gives a computable upper bound for $||b_\infty||$ with

$$||U_\infty|| = \max_{1 \le i \le \ell} \sup_{t \ge 0} |U_\infty^i(t)| \qquad (5.155)$$

where $U_\infty^i(t)$ denotes the ith input channel. Substituting in (5.150) now immediately yields the following, generally more conservative, alternative 'band' to that of (5.152)

$$Y_\infty^i(t) - \hat{\gamma}_k \le Y_k^i(t) \le Y_\infty^i(t) + \hat{\gamma}_k \qquad (5.156)$$

where

$$\hat{\gamma}_k := \frac{||L||^k ||L_b|| \ ||U_\infty||}{1 - ||L||} \qquad (5.157)$$

As one example, suppose that U_∞ is specified by (5.153). Then $||U_\infty|| = 1$ and this is one practically relevant case where (5.156) may prove computationally more attractive than (5.149). Finally, note that further consideration of the use of this computational information, together with that arising below from theorem 5.3.1, is postponed until the next chapter.

Note: The cited reference again details a number of potential refinements of the above analysis which do not require filtering operations.

Consider now the problem of obtaining bounds for the error $Y_k - Y_\infty$, $k \geq 0$, from theorem 5.3.1, where it is instructive to consider the unit memory case first. In which case, if Y_0 is again assumed to be zero for simplicity, theorem 5.3.1 states that

$$||Y_k - Y_\infty||_p \leq ||M_k||_p ||b_\infty||_p \tag{5.158}$$

where

$$||M_k||_p := (I_m - ||L||_p)^{-1} ||L||_p^k \tag{5.159}$$

and this matrix has been introduced for notational convenience.

Suppose also that $b_\infty(t)$ is again available from simulation studies on (5.151) with a given $U_\infty(t)$. Then it follows immediately that

$$||b_\infty||_p = [b_1, b_2, \ldots, b_m]^T \tag{5.160}$$

where

$$b_i := \sup_{t \geq 0} |b_\infty^i(t)|, \quad 1 \leq i \leq m \tag{5.161}$$

Suppose, therefore, that $||b_\infty||_p$ has been computed. Then it follows immediately from (5.158) that

$$||Y_k(t) - Y_\infty(t)||_p \leq ||Y_k - Y_\infty||_p \leq ||M_k||_p ||b_\infty||_p, \quad t \geq 0 \tag{5.162}$$

Equivalently, $Y_k^i(t)$, $t \geq 0$, lies in the 'band' defined by

$$Y_\infty^i(t) - m_k^i \leq Y_k^i(t) \leq Y_\infty^i(t) + m_k^i \tag{5.163}$$

where m_k^i denotes the ith entry in the m×1 column vector formed as the product of $||M_k||_p$ and $||b_\infty||_p$.

This band has the graphical interpretation shown in Figure 5.5 and is again obviously suitable for inclusion in a CAD environment. Further, its width from pass to pass is, in effect, governed by the relationship

$$||M_{k+1}||_p = ||M_k||_p ||L||_p, \quad k \geq 0 \tag{5.164}$$

In common with theorem 5.3.2, an alternative to (5.163) exists in terms of an upper bound on $||b_\infty||_p$ which, although more conservative in general, may prove computationally more attractive in certain cases. To develop this, first suppose again that the step response matrix, $W_b(t)$, of $b_\infty = L_b U_\infty$ is available. Then $||L_b||_p$ can be computed from $W_b(t)$ as per the computation of $||L||_p$ from $W^1(t)$ in section 5.2. Further,

$$||b_\infty||_p \leq ||L_b||_p ||U_\infty||_p \tag{5.165}$$

on use of (iv) of lemma 5.1.1 where

$$||U_\infty||_p = [U^1, U^2, \ldots, U^\ell]^T \tag{5.166}$$

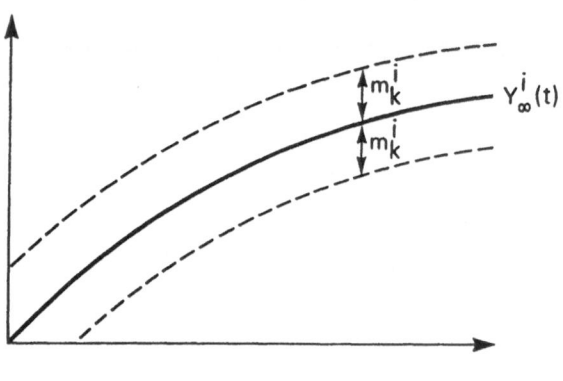

FIGURE 5.5

and

$$U^i := \sup_{t \geq 0} |U_\infty^i(t)| \qquad (5.167)$$

Hence

$$||M_k||_p||b_\infty||_p \leq ||M_k||_p||L_b||_p||U_\infty||_p := \tilde{m}_k \qquad (5.168)$$

and the following is a, generally more conservative, alternative 'band' to that of (5.163)

$$Y_\infty^i(t) - \tilde{m}_k^i \leq Y_k^i(t) \leq Y_\infty^i(t) + \tilde{m}_k^i \qquad (5.169)$$

where $\tilde{m}_k^i$ denotes the ith entry in $\tilde{m}_k$. Finally, if U_∞ is specified by (5.153) then

$$U^j = \begin{cases} 1, & i = j \\ \\ 0, & i \neq j \end{cases} \qquad (5.170)$$

and this is one practically relevant case where (5.169) may prove computationally more attractive than (5.163).

The generalisation of the analysis just completed to the application of theorem 5.3.1 in the non-unit memory case is a straightforward exercise. In particular, first note again that $S(E_\alpha, W_\alpha, L_\alpha)$ is defined by the 'companion form' based structure of (2.23)-(2.24) in this case. Further, $||L||_p$ is given by (5.48) and hence (5.159) translates to

$$||M_k||_p = (I_N - ||L||_p)^{-1}||L||_p^k \qquad (5.171)$$

Suppose also that $||M_k||_p$ is written in partitioned form as

$$||M_k||_p = \begin{bmatrix} ||M_k^1||_p & ||M_k^2||_p \\ \\ ||M_k^3||_p & ||M_k^4||_p \end{bmatrix} \qquad (5.172)$$

where $||M_k^4||_p$ is of dimension m×m. Then (5.162) translates to

$$||Y_k(t) - Y_\infty(t)||_p \leq ||Y_k - Y_\infty||_p \leq ||M_k^4||_p||b_\infty||_p, \quad t \geq 0 \qquad (5.173)$$

where $||b_\infty||_p$ is again given by (5.160)-(5.161). Based on this result, the analysis given above for the unit memory case generalises in a natural manner and hence the details are omitted.

Note: The roles of $Y_\infty^i(t)$ and $Y_k^i(t)$ in (5.149),(5.156),(5.163) and (5.169) can, of course, be reversed.

To conclude this section, the analysis below develops a refinement of the results given above on the error $Y_k - Y_\infty$, $k \geq 0$, based, essentially, on filtering the elements in the step response matrices of the associated conventional linear systems. Following this, the extension of all results developed in this section to the discrete non-unit memory process of example 2.3.7, or its unit memory version of example 2.3.8, is noted. In the former area, only the SISO unit memory case is considered since all others follow immediately as natural generalisations and the details can be found in the cited reference.

Consider first

$$X_\lambda(t_\lambda) := \{f: e^{\lambda t} f \in L_\infty(0, t_\lambda)\} \tag{5.174}$$

and take the norm of f as the norm of $e^{\lambda t} f$ in $L_\infty(0, t_\lambda)$. Then the following is the effective basis of the analysis given below for the SISO case and its natural generalisations.

Lemma 5.3.1: Suppose that L is a linear convolution operator mapping $L_\infty(0, +\infty)$ into itself with impulse response satisfying

$$|h_L(t)| \le h_o e^{-\gamma t}, \quad t \ge 0$$

for some $\gamma > 0$ and let Y_L denote its unit step response. Then for $t_\lambda < +\infty$ the induced operator norm of L restricted to $X_\lambda(t_\lambda)$ satisfies

$$||L|| \le N_{t_\lambda}(Y_L; \lambda), \quad \forall \lambda \tag{5.175}$$

where

$$N_{t_\lambda}(Y_L; \lambda) := |Y_L(0+)| + \sum_{k=1}^{N} e^{\beta_k(\lambda)} (N_{t_k}(Y_L) - N_{t_{k-1}}(Y_L)) \tag{5.176}$$

with

$$\beta_k(\lambda) = \begin{cases} \lambda\, t_k, & \lambda \ge 0 \\ \\ \lambda\, t_{k-1}, & \lambda < 0 \end{cases} \tag{5.177}$$

and $0 = t_o < t_1 < \ldots < t_N = t_\lambda$ is any partition of $[0, t_\lambda]$ with the property that $0 < h_1 \le t_k - t_{k-1} \le h_2 < +\infty$, $k \ge 1$. If $t_\lambda = +\infty$, L maps $X_\lambda(t_\lambda)$ into itself for $\lambda < \gamma$ with

$$||L|| \le N_\infty(Y_L; \lambda) < +\infty \tag{5.178}$$

Proof: The induced norm of L in $X_\lambda(t_\lambda)$ is

$$||L|| = |Y_L(0+)| + \int_0^{t_\lambda} e^{\lambda t} |h_L(t)| dt$$

$$\le |Y_L(0+)| + \sum_{k=1}^{N} e^{\beta_k(\lambda)} \int_{t_{k-1}}^{t_k} |h_L(t)| dt \tag{5.179}$$

which is just (5.176) and is clearly finite if $t_\lambda < +\infty$. If $t_\lambda = +\infty$ then $N = +\infty$ and if $\lambda \le 0$, $e^{\beta_k(\lambda)} \le 1$ and therefore

$$N_\infty(Y_L; \lambda) \le N_\infty(Y_L) < +\infty \tag{5.180}$$

Finally, if $0 < \lambda < \gamma$ then

$$e^{\beta_k(\lambda)} \int_{t_{k-1}}^{t_k} |h_L(t)| dt \leq h_0(t_k - t_{k-1}) e^{\lambda t_k - \gamma t_{k-1}}$$

$$\leq h_0(t_k - t_{k-1}) e^{(\lambda-\gamma)t_k + \gamma(t_k - t_{k-1})}$$

$$\leq h_0 h_2 e^{(\lambda-\gamma)kh_1} e^{\gamma h_2} \tag{5.181}$$

and the (infinite) series in (5.176) converges. ∎

It is easily verified that $N_{t_\lambda}(Y_L;\lambda)$ is continuous and monotonically increasing in both λ and t_λ with $N_0(Y_L;\lambda) = 0$ and $N_t(Y_L;0) = N_t(Y_L)$, i.e. the case of lemma 5.1.1 is recovered if $\lambda = 0$. Further, use of this lemma requires the availability of $\gamma > 0$ or, at least, the availability of appropriate information to determine the admissible range. Here it will be assumed, for simplicity, that this range is available in advance, with the cited reference giving full details of how to obtain it for a given example.

Return now to the SISO version of example 2.3.4 and suppose that the step response, $W^1(t)$, of the associated conventional linear system is available and satisfies assumption 5.2.2. Then this ensures that lemma 5.3.1 is applicable and use of (5.178) yields

$$||L|| \leq N_\infty(W^1;\lambda) \tag{5.182}$$

where $N_\infty(W^1;\lambda)$ is computed using (5.176). Further,

$$N_\infty(W^1;\lambda) < 1 \tag{5.183}$$

is a sufficient condition for stability along the pass and let $||.||_\lambda$ denote the norm in X_λ of (5.174), i.e.

$$||f||_\lambda = \sup_{t \geq 0} e^{\lambda t} |f(t)| \tag{5.184}$$

In which case it is easily shown by mirroring the proof of (5.144), i.e.

$$||Y_k - Y_\infty|| \leq ||L||^k \{||Y_0|| + \frac{||b_\infty||}{1 - ||L||}\}, \quad k \geq 0 \tag{5.185}$$

that

$$||Y_k - Y_\infty||_\lambda \leq ||L_\lambda||_\lambda^k \{||Y_0||_\lambda + \frac{||b_\infty||_\lambda}{1 - ||L_\lambda||_\lambda}\}, \quad k \geq 0 \tag{5.186}$$

where $||L_\lambda||_\lambda = N_\infty(W^1;\lambda)$. Hence, with Y_0 again assumed to be zero for simplicity,

$$e^{\lambda t} |Y_k(t) - Y_\infty(t)| \leq ||Y_k - Y_\infty||_\lambda \leq \hat{\gamma}_k, \quad t \geq 0 \tag{5.187}$$

where

$$\hat{\gamma}_k := \frac{||L_\lambda||_\lambda^k ||b_\infty||_\lambda}{1 - ||L_\lambda||_\lambda} \tag{5.188}$$

Suppose, therefore, that $||b_\infty||_\lambda$ is available, where its computation is an obvious extension of the analysis detailed earlier in this section for $||b_\infty||$ and hence the details are omitted. Then by (5.187), $Y_k(t)$, $t \geq 0$, lies in the 'band' defined by

$$|Y_k(t) - Y_\infty(t)| \leq \hat{\gamma}_k e^{-\lambda t} \tag{5.189}$$

or

$$Y_\infty(t) - \hat{\gamma}_k e^{-\lambda t} \leq Y_k(t) \leq Y_\infty(t) + \hat{\gamma}_k e^{-\lambda t} \tag{5.190}$$

This band has an obvious graphical interpretation and can easily be included within a CAD environment. If $\lambda = 0$, this band reduces to that of (5.149)-(5.150). Note also that the essential action of the exponential weighting, $e^{\lambda t}$, is to introduce a refinement which can, in principle, be used to tighten this band. In practice, it is envisaged that this band will be computed for a variety of choices of λ in a selection set. The cited reference contains complete details of this particular aspect together with the results of some very promising initial numerical studies. Note: The roles of $Y_\infty(t)$ and $Y_k(t)$ in (5.190) can, of course, be reversed.

As a final item in this section, return to the discrete non-unit memory linear repetitive process of example 2.3.7 or its unit memory version of example 2.3.8. Then it follows immediately that all of the results developed in this section extend in a natural manner to these processes where L_a^j, $1 \leq j \leq M$, is defined by (5.123) and b_∞ by (5.124) with $U_{k+1} = U_\infty$, $k \geq 0$. Hence the details are omitted except to note that all required total variation calculations are undertaken using lemma 5.1.2 or the discrete equivalent of lemma 5.3.1 as appropriate.

5.4 Interpass Smoothing

To date, no consideration has been given in this work to testing for stability in the presence of interpass smoothing effects. This is a common feature of a number of known industrial examples, such as long-wall coal cutting, and is, in effect, the name given to dynamic interaction which occurs between passes and distorts the previous pass profile(s). For example, the source of this interpass smoothing in the long-wall coal cutting case is the machine's weight (up to 5 tonnes) as it passes over.

Consider again the simulation-based stability tests of section 5.2. Then the purpose of this section is to consider the extension of these tests to one possible method of modelling interpass smoothing effects. This, in effect, assumes that the output at any point on the current pass is a function of the states and inputs at this point and of the complete profile on the previous pass. In particular, the case studied is those differential unit memory processes which can be described by the state-space model of example 2.3.5, i.e.

$$\dot{X}_{k+1}(t) = AX_{k+1}(t) + BU_{k+1}(t) + B_0 \int_0^\alpha K(t,\tau)Y_k(\tau)d\tau$$

$$Y_{k+1}(t) = CX_{k+1}(t)$$

$$0 \le t \le \alpha, \quad X_{k+1}(0) = 0, \quad k \ge 0 \tag{5.191}$$

Here the interpass interaction term $B_0 \int_0^\alpha K(t,\tau)Y_k(\tau)d\tau$ represents a 'smoothing out' of

the previous pass profile in a manner governed by the properties of the kernel $K(t,\tau)$. Note again that the particular choice of

$$K(t,\tau) = \delta(t-\tau)I_m \tag{5.192}$$

where δ denotes the Dirac delta function reduces (5.191) to the case of example 2.3.4. Further, the analysis presented below extends in a natural manner to the non-unit memory version of (5.191) (defined in an obvious manner) and to the corresponding discrete cases. Hence the details are omitted and can be found in the cited reference.

The effective basis for the analysis given below is the following result.

Lemma 5.4.1: Suppose that L is a bounded linear operator mapping $L_\infty(0,+\infty)$ into itself of the form

$$(LY)(t) = \int_0^\alpha K(t,\tau)Y_k(\tau)d\tau \tag{5.193}$$

Then

$$||L|| \le \sup_{t \ge 0} \int_0^\infty |K(t,\tau)|d\tau \tag{5.194}$$

and equality holds in (5.194) if $K \ge 0$, $\forall t, \tau$.

Proof: By definition

$$||L|| = \sup_{t \ge 0} \sup_{||Y||=1} |\int_0^\infty K(t,\tau)Y(\tau)d\tau|$$

$$\le \sup_{t \ge 0} \sup_{||Y||=1} \int_0^\infty |K(t,\tau)||Y(\tau)|d\tau$$

$$\le \sup_{t \ge 0} \sup_{||Y||=1} \int_0^\infty |K(t,\tau)|d\tau||Y||$$

$$= \sup_{t \ge 0} \int_0^\infty |K(t,\tau)|d\tau \tag{5.195}$$

Further, equality obviously holds in (5.195) if $K \ge 0$, $\forall t,\tau$. ∎

One choice of $K(t,\tau)$ is

$$K(t,\tau) = K_0 \, e^{-\beta|t-\tau|} \tag{5.196}$$

where K_0 and β are positive real scalars and $K(t,\tau) > 0$, $\forall t, \tau$. This represents so-called 'double sided exponential smoothing' and has been used, see the cited reference for full details, to develop a first practically realistic treatment of interpass smoothing effects in the long-wall coal cutting example. Further, it is easily shown that

$$||L|| = \frac{2K_0}{\beta} \tag{5.197}$$

in this case.

Return now to the state-space model of (5.191) and re-write it as

$$\dot{X}_{k+1}(t) = AX_{k+1}(t) + BU_{k+1}(t) + B_0V_k(t)$$

$$Y_{k+1}(t) = CX_{k+1}(t) \tag{5.198}$$

where

$$V_k(t) = \int_0^\alpha K(t,\tau)Y_k(\tau)d\tau \tag{5.199}$$

Then, proceeding formally, the Laplace transform description of the associated conventional linear system is given by

$$W^1(s) = G_1(s)V(s) \tag{5.200}$$

where $G_1(s)$ is (as before) the transfer-function matrix for the case of no interpass smoothing, and $V(s)$ is the Laplace transform of

$$V(t) = \int_0^\infty K(t,\tau)Y(\tau)d\tau := (KY)(t) \tag{5.201}$$

where K is the integral operator with kernel $K(t,\tau)$. It now follows that the L_∞ induced norm of G_1K is bounded above by

$$||L_s||_p = ||L||_p||\hat{K}||_p \tag{5.202}$$

where

$$\hat{K} := \sup_{t \geq 0} \int_0^\infty K(t,\tau)d\tau \tag{5.203}$$

(The supremum being interpreted with respect to the partial ordering).

The following result now provides a (computable) sufficient condition for stability along the pass of (5.191). Further, the proof of this result follows identical steps to that of theorem 5.2.1 and hence the details are omitted.

Theorem 5.4.1: Suppose that the matrix $||L_s||_p$ of (5.202) has been constructed for the differential unit memory linear repetitive process of (5.191). Then the extended linear repetitive process $S(E_\alpha,W_\alpha,L_\alpha)_{\alpha \geq \alpha_0}$ generated by this example with $\alpha \geq \alpha_0$ is stable along the pass if

$$r(||L_s||_p) < 1 \tag{5.204}$$

Note: As per section 5.2, $r(||L_s||_p) < 1$ can be replaced by $||L_s|| = ||(||L_s||_p)|| < 1$.

The cited reference contains the results of in depth theoretical and numerical studies which serve to confirm the potential of this means of representing interpass smoothing for certain cases of practical interest. As a sample of these, consider the SISO case where the unit step response of $G_1(s)$ is monotonic and sign definite and hence $||L||_p = |G_1(0)|$. Suppose also that the interpass kernel is given by (5.196) normalised such that

$$\int_0^\infty K(t,\tau)d\tau = 1 \tag{5.205}$$

and hence $\beta = K_0$. Consequently

$$||L_s||_p = ||L||_p \tag{5.206}$$

in this particular case, i.e. this (normalised) interpass smoothing has no effect on the stability condition of theorem 5.2.1 (or theorem 5.2.2).

As a final point, note that no results are yet available on the extension of the stability tests of chapter 4 to processes with interpass smoothing. Further, it is not immediately obvious how (if at all) this can be achieved. In particular, it may well be that the simulation-based approach of this chapter is the only feasible means of stability testing in this case. A definite conclusion to this conjecture, however, must await the outcome of much further research effort for which the results already available serve as an appropriate starting point.

Notes and References

The background material of section 5.1 is drawn from Owens and Chotai (1983) and the references therein. Smyth (1991) discusses the associated numerical computations. Theorem 5.2.1 of section 5.2 is from Rogers and Owens (1990 d) and theorem 5.2.2 is from Rogers and Owens (1990 e). For a comprehensive treatment of the filtering results of section 5.2 see Rogers and Owens (1990 f) and Rogers and Owens (1990 g) for the discrete versions of all of the results presented in section 5.2.

Theorem 5.3.1 and 5.3.2 of section 5.3 are from Rogers and Owens (1990 d) and Rogers and Owens (1990 e) respectively. The filtering results of this section are from Rogers and Owens (1990 f,g). Finally, the results of section 5.4 are from Rogers and Owens (1990 h) which also contains details of the extensions referred to in the text.

CHAPTER 6
CONTROLLER DESIGN - SOME INITIAL RESULTS

This chapter presents some initial work on controller design for the differential and discrete processes of examples 2.3.3 and 2.3.7. In particular, three control policies for these processes are formulated from practical considerations and feedback control schemes which use either state or output information are developed. Finally some candidate design algorithms are presented together with some systems theoretic properties, such as a return-difference matrix for the output feedback based schemes.

6.1 Control Policies and Feedback Control Schemes

By analogy with the conventional linear systems approach, consider the case of example 2.3.3 when there is 'no direct feedthrough' between input and output on any pass and hence the state-space model

$$\dot{X}^P_{k+1}(t) = A^P X^P_{k+1}(t) + B^P U_{k+1}(t) + \sum_{j=1}^{M} B^P_{j-1} Y_{k+1-j}(t)$$

$$Y_{k+1}(t) = C^P X^P_{k+1}(t) + \sum_{j=1}^{M} D^P_j Y_{k+1-j}(t)$$

$$X^P_{k+1}(t) \in R^{n_1}, \; Y_{k+1}(t) \in R^m, \; U_{k+1}(t) \in R^\ell$$

$$0 \leq t \leq \alpha, \; k \geq 0 \tag{6.1}$$

Then a study of industrial examples, such as bench mining systems, leads to the following three basic control policies. Note also that these policies extend in a natural manner to the corresponding discrete case of example 2.3.7. Hence the details, together with those corresponding to all other results presented in this chapter, are omitted.

Stability along the Pass - This is an obvious necessary item of any practically feasible control policy.

The Limit Profile Design Problem - Suppose that the particular example of (6.1) under consideration is asymptotically stable, i.e. theorem 3.1.4 holds. Then the origin of this control policy lies in the fact that the corresponding limit profile is described by

$$\dot{X}^P_\infty(t) = (A^P + \hat{B}^P(I_m - \hat{D}^P)^{-1}C^P)X^P_\infty(t) + B^P U_\infty(t)$$

$$Y_\infty(t) = (I_m - \hat{D}^P)^{-1}C^P X^P_\infty(t) \tag{6.2}$$

where

$$\hat{B}^P = \sum_{j=1}^{M} B^P_{j-1}, \; \hat{D}^P = \sum_{j=1}^{M} D^P_j \tag{6.3}$$

Hence, in effect, the repetitive dynamics in this case can, after a 'sufficiently large' number of passes, be described by a conventional linear systems state-space model. Further, stability along the pass implies that (6.2) is stable, i.e. all eigenvalues of the matrix $A^P + \hat{B}^P(I_m - \hat{D}^P)^{-1}C^P$ have strictly negative real parts. To see this, set $z = 1$ in the stability along the pass polynomial $A_p(s,z)$ of (3.116) and use (3.118).

Given stability along the pass and (6.2)-(6.3) it follows immediately that the repetitive systems behaviour after a 'sufficiently large' number of passes, formally the limit profile, can be classified in terms of well known conventional linear systems criteria. Further, in depth studies on a number of industrial examples has led to the following set of performance specifications which, in effect, constitute the limit profile design problem and which, for notational simplicity, is denoted by LPDP from this point onwards. The use of quotation marks indicates that the precise meaning of the terms within are a matter for judgement based on detailed knowledge of the particular application under consideration.

(i)   The process must be stable along the pass and hence the existence of a stable limit profile described by (6.2)-(6.3) is guaranteed. Further, the limit profile dynamics should satisfy such other additional conventional linear systems performance criteria as deemed appropriate. For example, the interaction effects in reponse to unit step demands should be within 'acceptable limits'.

(ii)   The output sequence $\{Y_k\}_{k \geq 1}$ must be within a 'specified bound', or band, of of Y_∞ after a fixed number of passes, say k^*, and remain within it for all successive passes $k > k^*$.

(iii)   The error $Y_k - Y_\infty$, $0 \leq k \leq k^*$, must be 'acceptable'.

Several variations of (i) - (iii) above exist and are detailed in the cited reference. Further, much work remains to be done on developing rules for refining the terms in quotation marks into design criteria which, where appropriate, should (ideally) display similar characteristics to existing well used conventional linear systems ones. Recall also that the simulation-based stability tests of chapter 5 yield, at no extra cost, computable information concerning (ii) and (iii) above in one special case of major practical interest. This particular aspect is considered again in section 6.3 which presents some initial results on the development of algorithms for designing the memoryless output feedback based schemes introduced below to solve this problem.

<u>The Repetitive Systems Disturbance Decoupling with Stability Problem</u> - This control policy again has its origins in industrial examples and is, in effect, based on regarding the previous pass profiles as disturbances. Its requirements can be summarised as follows and note that these obviously imply stability along the pass.

(i)   The pass profile $Y_k(t)$, $0 \leq t \leq \alpha$, should be independent of the pass profiles $Y_{k-j}(t)$, $0 \leq t \leq \alpha$, $1 \leq j \leq M$, for all passes $k \geq k^* \geq 1$ with an optimum choice of $k^* = 1$.

(ii)   Suppose that (i) above holds. Then for all passes $k \geq k^* \geq 1$ the systems dynamics are, in effect, described by the derived conventional linear system $L_D(A^P, B^P, C^P)$. (Simply delete the previous pass terms.) In effect, the limit profile is reached exactly on pass k^* with dynamics parameterised by the triple (A^P, B^P, C^P). Further, the dynamics of this limit profile should satisfy such conventional linear systems performance criteria as deemed appropriate. The minimum requirement here is stability and hence the choice of name for this control policy.

This problem will, again for notational simplicity, be denoted by RSDDSP from this point onwards. Further, it has well defined structural links with both the LPDP introduced above and its well known, and extensively researched, conventional linear systems counterpart. The details of this aspect can again be found in the cited reference. Section 6.4 presents some initial results of the development of algorithms for designing the feedback control schemes detailed below to solve this problem. One aspect of this is some initial results on the extension of the well known geometric based theory for the conventional linear systems case, with the eventual aim of providing synthesis type conditions for the existence of a solution.

It is important to note that the control policies defined above are by no means exhaustive. For example, a linear quadratic optimal control problem for (6.1) can be formulated which is the natural generalisation of its well known, and extensively researched, conventional linear systems counterpart. This, and other policies, are detailed in the cited references.

Consider now the problem of satisfying the various requirements of a given control policy for (6.1). Then, using the derived conventional linear systems case as motivation, one intuitively obvious approach is to use an appropriately defined feedback control scheme. This is the underlying theme of the remainder of this chapter for which the material below is essential background. In particular, some candidate schemes are defined and some relevant systems theoretic properties of these are developed.

The schemes presented below can be classified under the following two general headings.

(i)   Those which only explicitly use current pass information. These are termed current pass or memoryless.

(ii)   Those which explicitly use information from both the current and previous M pass profiles.

Further, all of these schemes are the natural generalisations of a corresponding scheme for the derived conventional linear system $L_D(A^P, B^P, C^P)$. In particular, they reduce to this scheme under the following conditions

(i)   Any previous pass terms are deleted.

(ii)   The pass subscript k + 1 is dropped.

(iii)   The concept of a pass length is ignored.

Note: In the remainder of this chapter, use of the term 'natural generalisation' should be interpreted as the result of applying (i)-(iii) above.

To introduce the first of these schemes, or laws, first note that a linear state feedback law for $L_D(A^P, B^P, C^P)$ has the structure

$$U(t) = FX^P(t) + GR(t) \qquad (6.4)$$

Here F and G are constant $\ell \times n_1$ and $\ell \times m$ matrices respectively to be selected and $R(t) \in R^m$ is a new external reference vector taken to represent desired behaviour. Further, (6.4) is a powerful and extensively studied control law where, for example, the cited text and the references therein give a comprehensive treatment of the known results. The natural generalisation of (6.4) for (6.1) is the so-called current pass, or memoryless, linear state feedback law

$$U_{k+1}(t) = FX^P_{k+1}(t) + GR_{k+1}(t), \ 0 \le t \le \alpha, \quad k \ge 0 \qquad (6.5)$$

Again F and G are $\ell \times n_1$ and $\ell \times m$ matrices respectively to be selected and $R_{k+1}(t)$ is a new external reference variable taken to represent desired behaviour on pass k + 1, k ≥ 0. Figure 6.1 shows a schematic diagram of this control law.

Substituting (6.5) into (6.1) yields the closed-loop state-space model

$$\dot{X}^P_{k+1}(t) = (A^P + B^P F)X^P_{k+1}(t) + B^P GR_{k+1}(t) + \sum_{j=1}^{M} B^P_{j-1} Y_{k+1-j}(t)$$

$$Y_{k+1}(t) = C^P X^P_{k+1}(t) + \sum_{j=1}^{M} D^P_j Y_{k+1-j}(t)$$

$$0 \le t \le \alpha, \ k \ge 0 \qquad (6.6)$$

This system is said to be closed since it has an identical structure to (6.1) and therefore necessary and sufficient conditions for closed-loop stability along the pass immediately result on interpreting theorem 3.3.7. Use of these, unlike the alternative set based on theorem 3.3.5, enable the computationally feasible tests of chapter 4 to be applied for a given F and G.

Two systems theoretic properties of (6.5) now immediately arise which are of fundamental underlying importance in terms of potential applications. The first of these is the fact that the matrices D^P_j, $1 \le j \le M$, of (6.1) and hence, by theorem 3.1.4, asymptotic stability is invariant under this control law. Further, it is clear that this is also true for all multipass causal, see (2.59) and Figure 2.7, feedback control schemes. This follows immediately from the following facts.

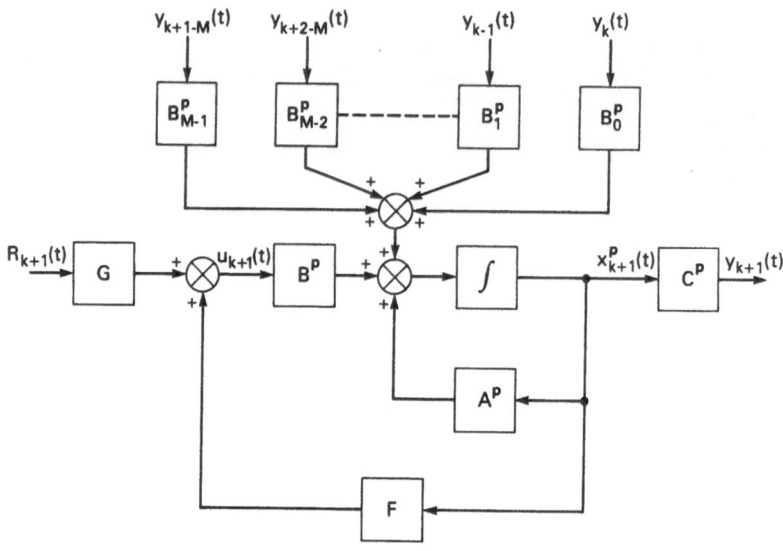

FIGURE 6.1

(i)   This system property only depends on the matrices D_j^P, $1 \leq j \leq M$.

(ii)   The output $Y_{k+1}(t)$ does not explicitly depend on the input $U_{k+1}(t)$, $0 \leq t \leq \alpha$, $k \geq 0$.

Suppose, however, that there is 'direct feedthrough' between input and output on any pass, i.e. the output equation of (6.1) reverts to that of example 2.3.3. Then in this case asymptotic stability is no longer invariant under a suitable choice of feedback control scheme. This subject is not considered here, however, and can be found in the cited reference.

Return now to the large sub-class of processes described by (6.1). Then the invariance property of asymptotic stability under multipass causal feedback control schemes has the following major implications

(i)   Any asymptotically unstable example cannot be stabilised by a multipass causal feedback control scheme.

(ii)   In systems theoretic terms, it is by no means clear at this stage how (if at all) this problem can be overcome. One approach may be to develop a more general representation which includes sub-classes such as examples 2.3.3 and 2.3.7 as special cases. No work has yet been undertaken in this general area, however, and it is left here as an open research problem.

In practical cases, however, it can be argued that asymptotic stability is always present due to the stabilising influence of resetting the initial conditions on each pass. To discuss this point further, consider, for simplicity, the unit memory version of (6.1) with $U_{k+1}(t) = 0$, $0 \leq t \leq \alpha$, $k \geq 0$, and zero state initial conditions on each pass. Then the initial ouptut on each pass is given by

$$Y_k(0) = D_1^k Y_0(0), \ k \geq 0 \tag{6.7}$$

and note again the condition of corollary 3.1.4 for asymptotic stability. Further, in physical terms, asymptotic stability requires that the initial output on each pass does not become unbounded as $k \to +\infty$. This will most certainly be the case in industrially orientated examples where the initial conditions on each pass are always finite. For example, in bench mining systems the cutting machine begins each pass from a fixed datum, or reference, level above the stone/coal interface.

Using the above results and observations, it can be concluded that the de-stabilising influences (if any) in industrially orientated cases are induced by the along the pass dynamics. Consequently the design studies in the remaining sections of this chapter will assume asymptotic stability. Note also that asymptotic stability always holds if $D_j^P = 0$, $1 \leq j \leq M$, i.e. no 'direct feedthrough' from previous pass profiles to the current one. Hence no loss of generality occurs if the terms arising from these matrices are deleted from the output equation of (6.1). This fact will also be exploited, where appropriate, in these design studies.

The second systems theoretic property of interest at this stage arises from the interpretation of (6.5) as the natural generalisation of (6.4). In particular, the closed-loop derived conventional linear system $L_D(A^P + B^P F, B^P G, C^P)$ is just the result of applying (6.4) to $L_D(A^P, B^P, C^P)$. Further, the standard design problem in this case is the choice of F for closed-loop stability, i.e. all eigenvalues of the matrix $A^P + B^P F$ have strictly negative real parts. This is the well known pole allocation, or assignment, problem and its basic form has a solution if, and only if, the pair $\{A^P, B^P\}$ is controllable. Given a desired set of locations in the open left-half of the complex plane for the eigenvalues of $A^P + B^P F$, the closed-loop poles, numerous algorithms exist for computing the corresponding F. One such set, for example, is based on the so-called controllable canonical form.

Suppose now that (6.1) is asymptotically stable. Then intepreting theorem 3.3.7 in terms of (6.6) immediately yields that stability of $L_D(A^P + B^P F, B^P G, C^P)$ is a necessary condition for closed-loop stability along the pass. Equivalently, the existence of a solution to the corresponding conventional linear systems problem is a necessary condition for closed-loop stability along the pass. As shown below, this result also holds for all other feedback control schemes introduced here, and is used in the next section to develop one candidate systematic procedure for designing any one of these schemes for closed-loop stability along the pass. Further, it provides a partial answer to the basic underlying synthesis problem of determining under what conditions (6.5) can be designed for closed-loop stability along the pass which is, as yet, unresolved in the general case. Note, however, that section 6.2 will also provide a complete answer to this question for certain sub-classes of (6.1).

Implementation of (6.5) requires measurement of all elements in the current pass state vector $X_{k+1}^P(t)$. By analogy with the conventional linear systems case, this may not be physically possible or practically feasible on, for example, financial grounds. In such cases, again by analogy with the conventional linear systems case, one possible option is to use a suitably designed observer or state estimation device. To date, however, no work has been undertaken on the development of an observer theory for (6.1) and this topic is left here as an open research problem. As an alternative, note that the output vector of (6.1) is available by assumption on each pass $k \geq 1$. Hence the material below follows the conventional linear systems case and introduces output feedback control schemes whose controllers explicitly use current, or a combination of current and previous, pass output information.

Consider, therefore, the output of (6.1) at time, or point, t on pass k, $k \geq 1$. Then the information in the following set is multipass causal (see also (2.59) and Figure 2.7) and can therefore be used for feedback control purposes.

$$Y := \{Y_k(\tau): 0 \leq \tau \leq t\} \cup \{Y_P(t): 0 \leq t \leq \alpha, \ 1-M \leq P \leq k-1\} \tag{6.8}$$

Clearly, however, the most appealing from an implementation standpoint will be those control schemes which explicitly use only information at point t on pass k since they will obviously have a simpler structure. This approach is followed below to produce one sub-class of so-called current pass, or memoryless, feedback control schemes. In particular, a so-called memoryless dynamic unity-negative feedback control scheme is developed which is the natural generalisation of its well known, and extensively used, conventional linear systems counterpart.

Suppose, therefore, that $R_{k+1}(t) \in R^m$ again denotes a new external reference vector taken to represent desired behaviour on pass k+1, $k \geq 0$. Further, define the so-called current pass-error vector as

$$e_{k+1}(t) = R_{k+1}(t) - Y_{k+1}(t), \quad 0 \leq t \leq \alpha, \quad k \geq 0 \tag{6.9}$$

Then a memoryless dynamic unity-negative feedback controller for (6.1) constructs the input $U_{k+1}(t)$, $k \geq 0$, as the output from

$$\dot{X}^c_{k+1}(t) = A^c X^c_{k+1}(t) + B^c e_{k+1}(t)$$
$$U_{k+1}(t) = C^c X^c_{k+1}(t) + D^c e_{k+1}(t)$$
$$0 \leq t \leq \alpha, \quad k \geq 0 \tag{6.10}$$

where $X^c_{k+1}(t) \in R^{n_2}$ denotes the internal state of (6.10). The resulting control scheme is shown in Figure 6.2 and it is clear that (6.9) and (6.10) describe a memoryless dynamic unity-negative feedback control scheme for (6.1). This is the natural generalisation of its conventional linear systems counterpart and (6.10) is again termed the forward-path controller. In effect, (6.9)-(6.10) is just the conventional linear systems scheme applied on pass $k + 1$, $k \geq 0$.

Specific choices of the matrices in (6.10) can now be made to generate a wide range of special cases which are the natural generalisations of their extensively used conventional linear systems counterparts. As one example, set $A^c = 0$, $B^c = 0$, $C^c = 0$ to yield

$$U_{k+1}(t) = D^c e_{k+1}(t), \quad 0 \leq t \leq \alpha, \quad k \geq 0 \tag{6.11}$$

In which case (6.9) and (6.11) describe a memoryless constant, or proportional, unity-negative feedback control scheme for (6.1).

To obtain the closed-loop state-space model, first define

$$X_{k+1}(t) = [X^P_{k+1}(t)^T, X^c_{k+1}(t)^T]^T \in R^n, \quad n = n_1 + n_2 \tag{6.12}$$

Then combining (6.1) and (6.10) yields the following composite state-space model describing the forward-path system

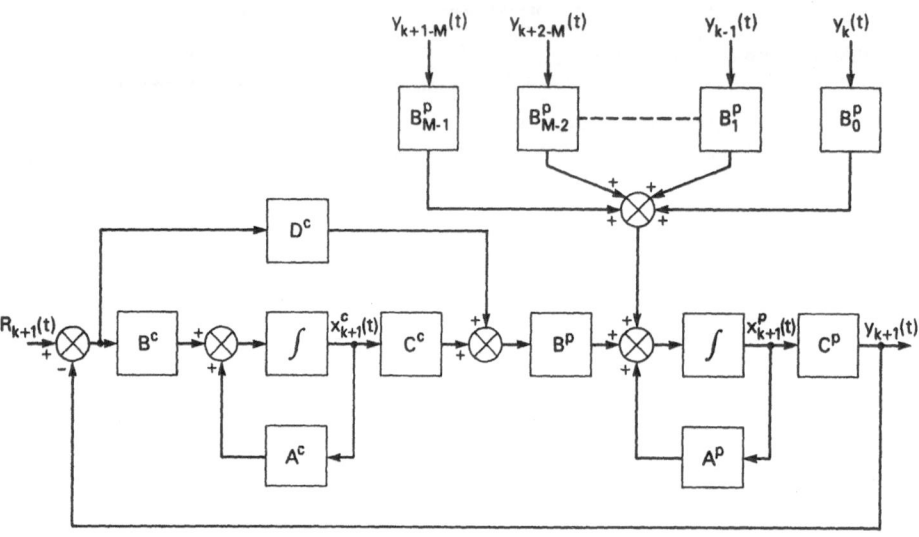

FIGURE 6.2

$$\dot{X}_{k+1}(t) = AX_{k+1}(t) + Be_{k+1}(t) + \sum_{j=1}^{M} B_{j-1}Y_{k+1-j}(t)$$

$$Y_{k+1}(t) = CX_{k+1}(t) + \sum_{j=1}^{M} D_j Y_{k+1-j}(t)$$

$$0 \leq t \leq \alpha, \quad k \geq 0 \tag{6.13}$$

where

$$A = \begin{bmatrix} A^P & B^P C^c \\ 0 & A^c \end{bmatrix}, \quad B = \begin{bmatrix} B^P D^c \\ B^c \end{bmatrix}, \quad B_{j-1} = \begin{bmatrix} B^P_{j-1} \\ 0 \end{bmatrix}, \quad 1 \leq j \leq M$$

$$C = [C^P \quad 0], \quad D_j = D^P_j, \; 1 \leq j \leq M \tag{6.14}$$

Further, combining (6.9) and (6.13)-(6.14) yields the closed-loop state-space model

$$\dot{X}_{k+1}(t) = (A-BC)X_{k+1}(t) + BR_{k+1}(t) + \sum_{j=1}^{M} (B_{j-1} - BD_j)Y_{k+1-j}(t)$$

$$Y_{k+1}(t) = CX_{k+1}(t) + \sum_{j=1}^{M} D_j Y_{k+1-j}(t)$$

$$0 \leq t \leq \alpha, \quad k \geq 0 \tag{6.15}$$

where

$$A - BC = \begin{bmatrix} A^P - B^P D^c C^P & B^P C^c \\ -B^c C^P & A^c \end{bmatrix} \tag{6.16}$$

Both (6.13) and (6.15) are closed in the sense that they have an identical structure to (6.1). Hence necessary and sufficient conditions for stability along the pass in both cases, which are computationally feasible to test, immediately result on appropriately interpreting theorem 3.3.7. Again, the matrices D^P_j, $1 \leq j \leq$ M, are invariant under this scheme and therefore, noting the conclusions based on (6.7), the analysis of later sections based on this scheme will assume open-loop asymptotic stability. Further, since this scheme is the natural generalisation of its conventional linear systems counterpart, it follows immediately that stability of the derived conventional linear system closed-loop is a necessary condition for stability along the pass of (6.15). Equivalently, the existence of a solution to the corresponding conventional linear systems design problem is a necessary condition for closed-loop stability along the pass. As in the current pass state feedback case, this result is used in the next section to develop one candidate systematic procedure for designing this scheme to give closed-loop stability along the pass. Note also that it provides a partial answer to the synthesis problem of determining the conditions under which this scheme can be designed for closed-loop stability along the pass.

The transfer-function matrix description plays a central role in the design of the conventional linear systems dynamic unity-negative feedback control scheme. In

particular, a number of currently available design techniques are based on the
so-called return-difference matrix defined in terms of the transfer-function matrix
of the forward-path system. The following analysis introduces, and develops a major
systems theoretic property of, the natural generalisation of the return-difference
matrix for memoryless dynamic unity-negative feedback control of (6.1).

First recall from section 2.5 that the 2D transfer-function matrix description
of (6.1) is

$$Y(s,z) = G^P(s,z)U(s,z) \tag{6.17}$$

where the $m \times \ell$ 2D transfer-function matrix $G^P(s,z)$ is given by

$$G^P(s,z) = (I_m - \sum_{j=1}^{M} G_j^P(s)z^{-j})^{-1}G_0^P(s) \tag{6.18}$$

with

$$G_0^P(s) = C^P(sI_{n_1} - A^P)^{-1}B^P \tag{6.19}$$

and

$$G_j^P(s) = C^P(sI_{n_1} - A^P)^{-1}B_{j-1}^P + D_j^P, \quad 1 \leq j \leq M \tag{6.20}$$

Further, consider the forward-path system (6.13)-(6.14) under the assumption of zero
state initial conditions on each pass and zero initial pass profiles. Then it is
easily shown that the 2D transfer-function matrix description of this system is given
by

$$\begin{aligned} Y(s,z) &= Q(s,z)e(s,z) \\ &= G^P(s,z)K(s,z)e(s,z) \end{aligned} \tag{6.21}$$

where $G^P(s,z)$ is defined by (6.18)-(6.20) and $K(s,z)$, the 2D transfer-function matrix
of the forward-path controller, is given by

$$K(s,z) \equiv G_0^C(s) = C^C(sI_{n_2} - A^C)^{-1}B^C + D^C \tag{6.22}$$

This result is the natural generalisation of its conventional linear systems
counterpart and states that $Q(s,z)$, the forward-path 2D transfer-function matrix, is
the product of the corresponding matrices for the plant and the controller. Finally,
substituting

$$e(s,z) = R(s,z) - Y(s,z) \tag{6.23}$$

into (6.21) and rearranging yields

$$Y(s,z) = H(s,z)R(s,z) \tag{6.24}$$

where the $m \times m$ 2D closed-loop transfer-function matrix $H(s,z)$ is defined by

$$H(s,z) = (I_m + Q(s,z))^{-1}Q(s,z) \tag{6.25}$$

The block diagram interpretation of (6.24) is shown in Figure 6.3 and this is
the natural generalisation of its conventional linear systems counterpart. Further,
it is a simple exercise to show that the block diagram algebra for the conventional

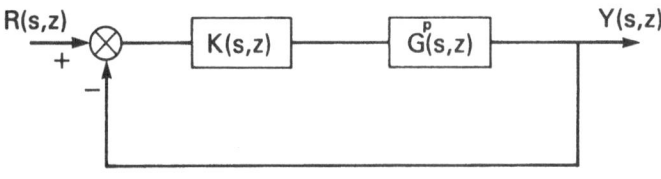

FIGURE 6.3

linear systems case also generalises in a natural manner. Hence the details are omitted.

In the case of the closed-loop derived conventional linear system, the return-difference matrix is defined by

$$T(s) = I_m + G_0^P(s)G_0^C(s) \tag{6.26}$$

Further, let $\rho_0(s)$ and $\rho_c(s)$ denote the open-loop forward-path and closed-loop characteristic polynomials respectively. Then the following result expresses closed-loop stability in terms of $T(s)$

$$\frac{\rho_c(s)}{\rho_0(s)} = |T(s)| \tag{6.27}$$

This return-difference relationship acts as the basis of a number of currently available design techniques (see, for example, the cited text and the references therein for comprehensive details of a representative cross-section), and the following analysis now develops the natural generalisation of this result for Figure 6.3.

The natural generalisation of (6.26) is

$$T(s,z) = I_m + G^P(s,z)K(s,z) \tag{6.28}$$

and in order to link this return-difference matrix to closed-loop stability along the pass it is first necessary to introduce a characteristic polynomial. This, in common with its conventional systems counterpart, should contain all the information necessary to determine stability along the pass. Consequently an obvious candidate for this open-loop is

$$\rho_0^P(s,z) = P_a^P(z)A_P^P(s,z) \tag{6.29}$$

where, from section 3.3, $P_a(z)$ and $A_P(s,z)$ are the asymptotic and stability along the pass polynomials respectively. In particular, from definitions 3.3.3 and 3.3.4 respectively,

$$P_a^P(z) = |Q^P(z)| = |I_m - z^{-1}D_1^P - \ldots - z^{-M}D_M^P| \tag{6.30}$$

and

$$A_P^P(s,z) = |sI_{n_1} - A^P - B^P(z)Q^P(z)^{-1}C^P| \tag{6.31}$$

where

$$B^P(z) = \sum_{j=1}^{M} B_{j-1}^P z^{-j} \tag{6.32}$$

Note also that $\rho_0^P(s,z)$ is the natural generalisation of its conventional linear systems counterpart $\rho_0^P(s) = |sI_{n_1} - A^P|$. Further, by Schur's formula

$$\rho_0^P(s,z) = \begin{vmatrix} sI_{n_1} - A^P & -B^P(z) \\ & \\ -C^P & Q^P(z) \end{vmatrix} \qquad (6.33)$$

and define the sets $\bar{D}$ and $\bar{U}$ by

$$\bar{D} = \{s: \text{Re}\{s\} \geq 0\} \qquad (6.34)$$

and

$$\bar{U} = \{z: |z| \geq 1\} \qquad (6.35)$$

respectively. Then the following result characterises stability along the pass of (6.1) in terms of its characteristic polynomial.

Theorem 6.1.1: With the assumptions of theorem 3.3.5 (interpreted in terms of (6.1)) the extended linear repetitive process $S(E_\alpha, W_\alpha, L_\alpha)_{\alpha \geq \alpha_0}$ generated by (6.1) with $\alpha \geq \alpha_0$ is stable along the pass if, and only if,

$$\rho^P(s,z) \neq 0 \text{ in } \bar{D} \times \bar{U} \qquad (6.36)$$

Proof: This, in effect, consists of showing that (6.36) is equivalent to the conditions of theorem 3.3.7 which are, in turn, equivalent to those of theorem 3.3.5.

Suppose first, therefore, that (6.36) holds and hence by Schur's formula

$$\rho_0^P(s,z) = |sI_{n_1} - A^P| |Q^P(z) - C^P(sI_{n_1} - A^P)^{-1} B^P(z)|$$

$$= \frac{|sI_{n_1} - A^P| \; |zI_N - G^P(s)|}{z^N} \neq 0 \text{ in } \bar{D} \times \bar{U} \qquad (6.37)$$

where $G^P(s)$ is the interpass transfer-function matrix of (2.90) for (6.1). Equivalently,

$$|sI_{n_1} - A^P| \neq 0 \text{ in } \bar{D} \qquad (6.38)$$

and

$$|zI_N - G^P(s)| \neq 0 \text{ in } \bar{D} \times \bar{U} \qquad (6.39)$$

Condition (b) of theorem 3.3.7 is now immediate from (6.38) and setting $s = i\omega$ in (6.39) generates (a) and (c) of this same result.

Conversely, suppose that (a)-(c) of theorem 3.3.7 hold. Then following the second part of the proof of this result immediately yields

$$\rho_0^P(s,z) = P_a^P(z) A_p^P(s,z) \neq 0 \text{ in } \bar{D} \times \bar{U} \qquad (6.40)$$

and the proof is complete. ∎

One immediate use of this characteristic polynomial is to introduce the concept of a pole in terms of the solutions of $\rho_0^P(s,z) = 0$. This is again the natural generalisation of its conventional linear systems counterpart and has a well defined physical interpretation which can be used to provide a physically based explanation of the differences between asymptotic stability and stability along the pass. Computation of the poles of (6.1) is not a feasible proposition, however, and hence they are not considered further in this work. The details for the unit memory case can be found in the cited reference and it should also be noted that the extension of some of the results given there to non-unit memory processes is still an open research question.

Return now to $T(s,z)$ of (6.28) and let $\rho_c(s,z)$ and $\rho_0(s,z)$ denote the closed-loop and open-loop characteristic polynomials, i.e.

$$\rho_c(s,z) = \begin{vmatrix} sI_n - A+BC & -(B(z) - BD(z)) \\ -C & Q(z) \end{vmatrix} \tag{6.41}$$

and

$$\rho_0(s,z) = \begin{vmatrix} sI_n - A & -B(z) \\ -C & Q(z) \end{vmatrix} \tag{6.42}$$

respectively. Then the following result is the natural generalisation of its conventional linear systems counterpart and links this return-difference matrix to closed-loop stability along the pass.

Theorem 6.1.2: The return-difference matrix for the differential non-unit memory linear repetitive process (6.1) under memoryless dynamic unity-negative feedback control satisfies

$$\frac{\rho_c(s,z)}{\rho_0(s,z)} = |T(s,z)| \tag{6.43}$$

Proof: This follows immediately on use of Schur's formula and appropriate results from the theory of determinants. Hence the details are omitted. ∎

Given that it is the natural generalisation, it can be conjectured that $T(s,z)$ should play a similar role to its conventional linear systems counterpart in terms of the design of this scheme for, say, closed-loop stability along the pass. No work has yet been undertaken in this area, however, and it is left here as an open research problem. Instead, the design studies of the next section, in particular, will develop a systematic controller design procedure based on the fact that a solution to the derived conventional linear systems problem is a necessary condition for closed-loop stability along the pass. Further, this procedure can be implemented using only standard conventional linear systems techniques such as the characteristic locus.

The concept of a memoryless dynamic unity-negative feedback control scheme for (6.1) is easily extended to include a memoryless dynamic (i.e. non-unity) feedback loop and/or memoryless minor loop compensation. The most general case is shown in Figure 6.4 where the notation L.R.P denotes (6.1) and M.L.R.P. denotes a process of the form (6.10). Further, it is easily shown that all relevant elements of the analysis given to date in this section extend in a natural manner to these cases. Hence the details are omitted and can be found in the cited reference. Note also

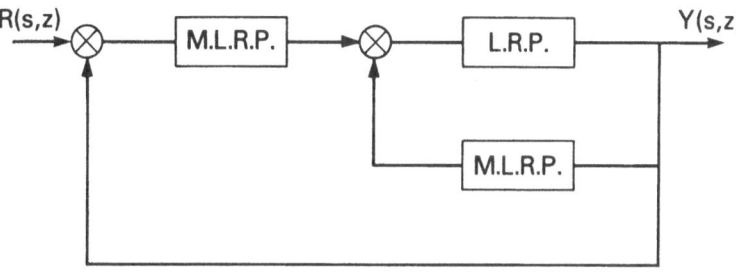

FIGURE 6.4

that these schemes by no means exhaust the possibilities based on memoryless
information and the cited reference again contains detailed information on this
point.

Suppose now that a case exists where the objectives cannot be met by a
permissible memoryless scheme. Then one obvious option in such cases is to consider
the use of controllers with memory, i.e. schemes where some, or all, of the
controllers explicitly use information from some, or all, of the previous pass
components of the causal set (6.8). Of all the various possibilities, the analysis
of section 6.4 makes particular use of so-called proportional repetitive minor loop
compensation schemes. Consequently only these schemes are detailed here with the
cited reference giving a comprehensive treatment of all other possibilities.

Consider again, therefore, the process of (6.1). Then a current pass, or
memoryless, linear state feedback law for this case with proportional repetitive
minor loop compensation has the structure.

$$U_{k+1}(t) = FX^P_{k+1}(t) + GR_{k+1}(t) - \sum_{j=1}^{M} K_j Y_{k+1-j}(t), \; 0 \le t \le \alpha, \; k \ge 0 \quad (6.44)$$

Here F,G and K_j, $1 \le j \le M$, are $\ell \times n_1$, $\ell \times m$ and $\ell \times m$ matrices respectively to be
selected and $R_{k+1}(t)$ is again a new external reference vector taken to represent
desired behaviour on pass k+1, $k \ge 0$. Figure 6.5 shows a schematic diagram of this
control law. Note also that it reduces to (6.5) if $K_j = 0$, $1 \le j \le M$, i.e. the
previous pass terms are deleted.

Substituting (6.44) into (6.1) yields the closed-loop state-space model

$$\dot{X}^P_{k+1}(t) = (A^P + B^P F)X^P_{k+1}(t) + B^P GR_{k+1}(t) + \sum_{j=1}^{M} (B^P_{j-1} - B^P K_j)Y_{k+1-j}(t)$$

$$Y_{k+1}(t) = C^P X^P_{k+1}(t) + \sum_{j=1}^{M} D^P_j Y_{k+1-j}(t)$$

$$0 \le t \le \alpha, \; k \ge 0 \quad\quad\quad (6.45)$$

Further, (6.45) is closed in the sense that it has an identical structure to (6.1).
Hence necessary and sufficient conditions for stability along the pass closed-loop,
which are computationally feasible to test, immediately result on interpreting

theorem 3.3.7. Again, the matrices D^P_j, $1 \le j \le M$, are invariant and hence, using the
conclusions of the analysis based on (6.6)-(6.7), it is necessary to assume open-loop
asymptotic stability.

The extra design freedom in this control law is clearly the matrices K_j,
$1 \le j \le M$. Note also that these matrices only affect the previous pass driving terms
in the state equation and hence only the interpretation of condition (c) of theorem
3.3.7 for closed-loop stability along the pass. In particular, stability of the

derived system $L_D(A^P + B^P F, B^P G, C^P)$ is again a necessary condition for closed-loop
stability along the pass. Further, in common with the case of (6.5), this fact can,
see the cited reference for complete details, be used to develop a candidate
systematic procedure for design to ensure, say, closed-loop stability along the pass.

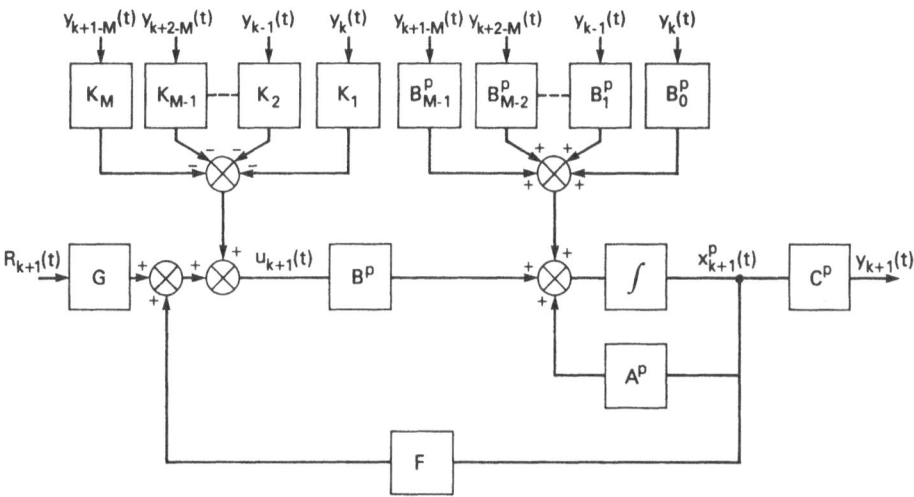

FIGURE 6.5

This procedure requires rules for choosing the K_j, $1 \leq j \leq M$, i.e. the repetitive minor loop compensation, and in all but a number of special cases, such as the one detailed in section 6.4 for the scheme introduced below, this is still very much an open research problem.

Returning now to the case of output feedback based schemes, suppose that (6.9) again defines the current pass error vector. Then a memoryless dynamic unity-negative feedback controller with proportional repetitive minor loop compensation for (6.1) constructs the input $U_{k+1}(t)$, $k \geq 0$, as

$$U_{k+1}(t) = Y^C_{k+1}(t) - \sum_{j=1}^{M} K_j Y_{k+1-j}(t), \quad 0 \leq t \leq \alpha, \ k \geq 0 \tag{6.46}$$

where the K_j, $1 \leq j \leq M$, are $\ell \times m$ matrices and $Y^C_{k+1}(t)$ is the output from

$$\dot{X}^C_{k+1}(t) = A^C X^C_{k+1}(t) + B^C e_{k+1}(t)$$
$$Y^C_{k+1}(t) = C^C X^C_{k+1}(t) + D^C e_{k+1}(t)$$
$$0 \leq t \leq \alpha, \quad k \geq 0 \tag{6.47}$$

where $X^C_{k+1}(t) \in R^{n_2}$ denotes the internal state of (6.47). Note also that this scheme reduces to that described by (6.9) - (6.10) if $K_j = 0$, $1 \leq j \leq M$, i.e. the previous pass terms are deleted.

To obtain the closed-loop state-space model, first define $X_{k+1}(t) \in R^n$ as in (6.12). Then combining (6.1) and (6.46) - (6.47) yields the following composite state-space model describing the forward-path system

$$\dot{X}_{k+1}(t) = A X_{k+1}(t) + B e_{k+1}(t) + \sum_{j=1}^{M} B_{j-1} Y_{k+1-j}(t)$$
$$Y_{k+1}(t) = C X_{k+1}(t) + \sum_{j=1}^{M} D_j Y_{k+1-j}(t)$$
$$0 \leq t \leq \alpha, \quad k \geq 0 \tag{6.48}$$

where A, B, C and D_j, $1 \leq j \leq M$, are again given by (6.14) but here

$$B_{j-1} = \begin{bmatrix} B^P_{j-1} - B^P K_j \\ 0 \end{bmatrix}, \quad 1 \leq j \leq M \tag{6.49}$$

Further, combining (6.9) and (6.48) - (6.49) yields the closed-loop state-space model

$$\dot{X}_{k+1}(t) = (A - BC)X_{k+1}(t) + BR_{k+1}(t) + \sum_{j=1}^{M}(B_{j-1} - BD_j)Y_{k+1-j}(t)$$
$$Y_{k+1}(t) = C X_{k+1}(t) + \sum_{j=1}^{M} D_j Y_{k+1-j}(t)$$
$$0 \leq t \leq \alpha, \quad k \geq 0 \tag{6.50}$$

where $A-BC$ is again given by (6.16).

Both (6.48) and (6.50) are closed in the sense that they have an identical structure to (6.1). Hence necessary and sufficient conditions for stability along the pass in both cases, which are computationally feasible to test, immediately result on appropriately interpreting theorem 3.3.7. Again, the matrices D^P_j, $1 \leq j \leq M$, are invariant and hence, see (6.6) - (6.7), it is necessary to assume open-loop asymptotic stability.

As with its state feedback based counterpart of (6.44), the extra design freedom in (6.46) - (6.47) is clearly the matrices K_j, $1 \leq j \leq M$, and these matrices again only affect the previous pass driving terms in the state equation. Hence they only influence the interpretation of condition (c) of theorem 3.3.7 closed-loop and stability of the derived system $L_D(A-BC,B,C)$ is a necessary condition for closed-loop stability along the pass. The cited reference shows how this fact can be exploited to develop one candidate systematic procedure for design to ensure, say, closed-loop stability along the pass. Further, section 6.4 uses this scheme to solve the RSDDSP for one sub-class of (6.1) which is of industrial interest. This analysis also yields some initial results on the development (see also the discussion after (6.45)) of rules for effectively selecting the repetitive minor loop compensation.

As an alternative to the state-space approach detailed above, the scheme defined by (6.46) - (6.47) can be described in 2D transfer-function matrix terms. Further, all results for this description follow immediately as straightforward extensions of their counterparts for the scheme of Figure 6.3. Hence only the final forms of the basic underlying results are stated here. The first of these is the closed-loop 2D transfer-function matrix description whose block diagram interpretation is shown in Figure 6.6

$$Y(s,z) = H(s,z)R(s,z) \tag{6.51}$$

where the $m \times m$ 2D closed-loop transfer-function matrix $H(s,z)$ is defined by

$$H(s,z) = (I_m + Q(s,z))^{-1}Q(s,z) \tag{6.52}$$

with

$$Q(s,z) = (I_m + G^P(s,z)K_L(s,z))^{-1}G^P(s,z)K(s,z) \tag{6.53}$$

Here $G^P(s,z)$ is the 2D transfer-function matrix of (6.1) defined by (6.18) - (6.20), $K(s,z)$ is defined by (6.22), and

$$K_L(s,z) \equiv K_L(z) = \sum_{j=1}^{M} K_j z^{-j} \tag{6.54}$$

Suppose also that the return-difference matrix for this case is defined as

$$T(s,z) = I_m + Q(s,z) \tag{6.55}$$

Then the result of theorem 6.1.2 still holds with $\rho_0(s,z)$ and $\rho_c(s,z)$ defined in terms of (6.48) and (6.50) respectively.

As a final point in this section, note that the schemes detailed or referred to here by no means exhaust the possibilities for controlling (6.1), or its discrete counterpart, based on causal memoryless and/or non-memoryless information. This point is considered in detail in the cited reference which, for example, introduces schemes which include 'feedforward' elements. In terms of this work, however, the schemes detailed in this section are sufficient to demonstrate the potential power of

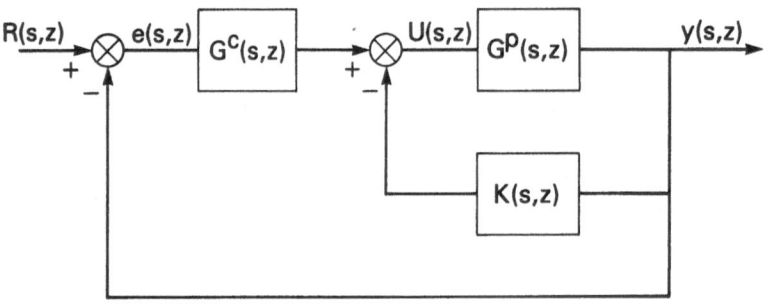

FIGURE 6.6

appropriately specified feedback control schemes in regulating the behaviour of differential (and discrete) non-unit memory linear repetitive processes.

An attractive feature of these schemes from an applications standpoint is their relative simplicity and hence the possibility of implementation without recourse to special purpose hardware/software. This is particularly true for the memoryless schemes which, see the cited reference for a complete treatment, have the simplest structure in terms of the the information to be logged and/or stored in order to actuate the controller. Further, this point strongly suggests that the potential of these schemes should be fully evaluated before recourse to other schemes with a more complex structure.

6.2 Stability Along the Pass

This section presents some initial results on the design of the schemes introduced in the previous section for closed-loop stability along the pass. In effect, the results presented consist of one candidate design algorithm plus an in-depth treatment of its application to certain sub-classes and details of some possible future research topics. These results represent the first output on this property, which is an essential necessary item of any practically feasible control policy and only apply (directly) to the memoryless schemes. (Consider again the discussion at the end of the last section on the 'complexity' of the schemes detailed there.)

An important result from section 6.1 is that asymptotic stability of (6.1) is invariant under all multipass causal feedback control schemes. Further, it was argued there that this property is always present in practical cases due to the stabilising influence of resetting the initial conditions on each pass. Consequently the analysis of this section assumes that (6.1) is asymptotically stable. Note again that asymptotic stability always holds if $D_j^P = 0$, $1 \leq j \leq M$, and hence no loss of generality results from deleting the previous pass terms from the output equation of (6.1).

Suppose, therefore that (6.1) is asymptotically stable and consider an arbitrary memoryless scheme from those introduced in section 6.1. Then interpreting theorem 3.3.7 and, in particular, conditions (b) and (c) of this result now gives necessary and sufficient conditions for closed-loop stability along the pass. Further, this scheme is the natural generalisation of its conventional linear systems counterpart. Hence, as also discussed briefly in section 6.1, the existence of a solution to the corresponding conventional linear systems problem is a necessary condition for closed-loop stability along the pass. This fact leads immediately to the following candidate systematic procedure, or design algorithm, for solving the repetitive systems problem which assumes that the derived conventional linear systems problem has a solution. For example, in the case of linear state feedback this assumption, in effect, requires that the pair $\{A^P, B^P\}$ is controllable.

<u>STEP 1</u>: Solve the derived conventional linear systems problem using any appropriate design algorithm.

<u>STEP 2</u>: Interpret condition (c) of theorem 3.3.7 in terms of the closed-loop system resulting from application of the controller designed at step 1. Then proceed to test this condition by applying the appropriate step in either of the systematic procedures developed in section 4.2. On completion, the following two options exist.

<u>STEP 3</u>: If step 2 holds the closed-loop system is stable along the pass. In which case proceed (if required) to examine other performance specifications such as the rate of approach to the limit profile.

<u>STEP 4</u>: The closed-loop system is unstable along the pass if step 2 does not hold. In which case return to step 1 and re-design (if possible).

In effect, all steps in the above procedure can be completed using standard conventional linear systems techniques. Suppose also that step 1 is completed using a technique, for example the characteristic locus or inverse Nyquist array in the case of the memoryless dynamic unity-negative feedback control scheme, for which well tested software is already available. Further, complete step 2 using the first systematic procedure of section 4.2. Then the above procedure is clearly suitable for a CAD environment. This is the subject of on-going research, see the cited reference for a complete update, which is developing the necessary infrastructure for inclusion in a user friendly interactive package. Coupled with this, research is proceeding in the following two general areas which, in effect, determine the applicable range of these memoryless schemes and their efficiency in terms of, for example, computational cost of the above design procedure.

(i) The classification of the sub-classes (if any) of (6.1) for which this problem can be solved by a particular scheme. Ideally what is required here are 'synthesis type' results similar to those for certain of the derived conventional linear systems problems, such as linear state feedback. This is termed the existence problem.

(ii) The development of 'easy to use' design rules such as relative stability and/or performance indicators. These can then be incorporated into the interactive package to give increased computational efficiency. This is an important aspect since, on the assumption that a solution exists, it may require more than one iteration of the above procedure (or any alternative) to yield a successful controller.

All controller parameters appear in the closed-loop derived system of the memoryless schemes under consideration here. Further, stability of this derived system is a necessary condition for closed-loop stability along the pass. Equivalently, a solution to the derived problem is also necessary for a solution to the repetitive systems problem. Hence a partial answer to the existence problem already exists in all these cases.

The development of a complete solution to this problem for a number of the potentially more common schemes, such as state feedback and unity-negative output feedback control, is the subject of an on-going research programme. This has already yielded solutions for a number of sub-classes characterised by certain structural properties and, see the cited references for complete details, provided useful pointers for more general cases. To illustrate the currently available results, and demonstrate the application of the above procedure, the following two cases are now detailed.

Consider first, therefore, the SISO unit memory case where the derived and associated conventional linear systems are given by $L_D(A^P, B^P, C^P)$ and $L_A^1(A^P, k_0 B^P, C^P)$ respectively, where $k_0 \neq 0$ is a positive real scalar. In which case it is easily shown that these systems have the same zeros which are assumed to be real, negative and distinct, i.e. a special case of the minimum phase property. Suppose also that $C^P B^P \neq 0$ or, equivalently, these systems have n poles and q = n - 1 zeros. Then application of the above procedure to design a current pass state feedback law under these assumptions proceeds as follows.

STEP 1: The derived conventional linear systems problem has a solution under the assumption that the pair $\{A^P, B^P\}$ is controllable. In which case any one of numerous well documented algorithms can be used to compute the row vector F which assigns the eigenvalues of $A^P + B^P F$ to a specified set of locations λ_i, $1 \leq i \leq n$, in the open left-half of the complex plane.

STEP 2: The closed-loop interpass transfer-function, denoted $G_c(s)$, is given by

$$G_c(s) = \frac{k_0 \prod_{i-1}^{q} (s - z_i)}{\prod_{i=1}^{n} (s - \lambda_i)} \tag{6.56}$$

where the z_i are the zeros of $L_D(A^P, B^P, C^P)$ which are real, distinct and negative. Hence stability along the pass if, and only if, the frequency response plot generated by $G_c(i\omega)$, $\forall$ real $\omega \geq 0$, lies entirely within the unit circle in the complex plane. Suppose also that the λ_i are assigned to locations γ_i which are real, distinct and negative and satisfy

$$\gamma_1 < z_1 < \gamma_2 < \ldots < z_{n-1} < \gamma_n \tag{6.57}$$

Then, using the analysis of lemma 5.2.3 and example 5.2.1, this special case is stable along the pass if, and only if,

$$|k_0| |\prod_{i=1}^{q} z_i| < |\prod_{i=1}^{n} \lambda_i| \tag{6.58}$$

or, equivalently, the maximum value of the frequency response occurs at $\omega = 0$.

Note 1: The condition of (6.58) can always be satisfied by placing γ_1 'far enough' to the left of the origin on the real line.

<u>Note 2:</u> The results presented here form part of the analysis for a sub-class of
(6.1) with so-called 'fading memory' which can be found in the cited reference.

The second case detailed here is where the derived system has the structure of a
multivariable first order lag. In particular, suppose that $m = \ell = n_1$ and consider,
for simplicity, the unit memory case. Then, after use of a state transformation, if
necessary, the state-space model can be written as

$$\dot{X}_{k+1}(t) = - A_0^{-1} A_1 X_{k+1}(t) + A_0^{-1} U_{k+1}(t) + B_0 Y_k(t)$$

$$Y_{k+1}(t) = I_m X_{k+1}(t)$$

$$0 \leq t \leq \alpha, \quad k \geq 0 \tag{6.59}$$

where A_0 and A_1 are real constant matrices. Further, application of the above
procedure to design a memoryless proportional unity-negative feedback control scheme
for this case proceeds as follows.

<u>STEP 1:</u> Select the forward-path controller as

$$K = \rho A_0 - A_1 \tag{6.60}$$

where ρ is a positive real scalar. Then using the theory of a multivariable first
order lag (see the cited reference for complete details) it is easily shown that the
closed-loop derived system is stable $\forall \rho$

<u>STEP 2:</u> The closed-loop interpass transfer-function matrix is given by

$$G_c(s) = (I_m + G_0^P(s)K)^{-1} G_1^P(s)$$
$$= \frac{1}{s+\rho} B_0 \tag{6.61}$$

Further, it follows immediately on examining (6.61) that condition (c) of theorem
3.3.7 holds closed-loop if, and only if,

$$\rho > r(B_0) \tag{6.62}$$

where $r(.)$ again denotes the spectral radius. Finally, combining (6.62) with the
result of step 1 yields closed-loop stability along the pass $\forall \rho > r(B_0)$ which can
always be satisfied by choosing a 'high enough' value of ρ.

<u>Note:</u> The cited reference gives the generalisation of this analysis to the non-unit
memory case, which is a straightforward exercise.

Suppose now that a solution exists for the particular scheme under
consideration. Then, as noted under (ii) above, it may require more than one
iteration of the above procedure (or any alternative) to yield a successful
controller with the consequent prospect of a heavy computational load. Hence the
development of 'easy to use' design rules such as relative stability and/or
performance indicators is one obvious route to obtaining maximum computational
efficiency. This general area is the subject of an on-going research programme which
is being undertaken in parallel with the development of the interactive package. The

details of this are not discussed here and a comprehensive treatment of progress to-date can again be found in the cited reference.

In summary, therefore, this section has considered the design of an arbitrary memoryless scheme from those introduced in the previous section for closed-loop stability along the pass. The end product has been a systematic design procedure for solving this problem which, in effect, can be completed using standard conventional linear systems schemes and is suitable for a CAD environment. Further, an in-depth treatment of its application to two sub-classes of (6.1) has been presented. This has given a complete solution to the fundamental underlying existence problem for these cases and has also highlighted other possible future research topics.

At this stage, it is not possible to fully assess the potential of this procedure, or the memoryless schemes of section (6.1), in terms of solving this most basic of control problems. In particular, much work remains to be done in the areas outlined here, and others detailed in the cited reference, before this question can be realistically considered. The progress to date, coupled with the fact (refer again to the discussion at the end of the previous section) that these memoryless schemes have the simplest structure in implementation terms, strongly suggests that this procedure should be fully evaluated before considering other possibilities. These possibilities include the use of the sufficient, but not necessary, simulation-based stability tests of chapter 5 as a basis and the cited reference gives further information on this particular point.

6.3 The Limit Profile Design Problem

This section presents some initial results on the design of the schemes introduced in section 6.1 to solve the LPDP. The format is similar to section 6.2 in that only the memoryless schemes are considered to produce the first output on this problem. In effect, the results presented consist of one candidate design procedure plus an in-depth treatment of its application to one sub-class and some possible future research topics. Further, these results are based on use of the simulation-based stability tests of chapter 5 since these yield, at no extra cost, unique computable information concerning two components of this problem in one special case of major practical interest.

Consider, therefore, (6.1) with $D_j^p = 0$, $1 \leq j \leq M$, for simplicity, and hence the corresponding limit profile is described by the state-space model of (6.2)-(6.3) with $\hat{D}^p = 0$ or, in transfer-function matrix terms, by

$$Y_\infty(s) = G_\infty^P(s) U_\infty(s) \tag{6.63}$$

where

$$G_\infty^P(s) = C^P(sI_{n_1} - A^P - B^P C^P)^{-1} \hat{B}^P \tag{6.64}$$

Further, as a representative choice, consider the memoryless dynamic unity-negative feedback control scheme introduced in section 6.1. Then here the control action on the limit profile is described in state-space terms (replace all variables in (6.9) and (6.10) by their strong limits) by

$$\dot{X}_\infty^c(t) = A^c X_\infty^c(t) + B^c e_\infty(t)$$

$$U_\infty(t) = C^c X_\infty^c(t) + D^c e_\infty(t)$$

$$0 \leq t \leq \alpha \tag{6.65}$$

where

$$e_\infty(t) = R_\infty(t) - Y_\infty(t) \tag{6.66}$$

or, in transfer-function matrix terms, by

$$U_\infty(s) = G_0^c(s) e_\infty(s) \tag{6.67}$$

where $G_0^c(s)$ is given by (6.22). Hence the closed-loop limit profile can be regarded as the conventional linear systems unity-negative feedback control scheme for $G_\infty^P(s)$ with forward-path controller dynamics defined by $G_0^c(s)$. Suppose also that the closed-loop specifications for the limit profile can be achieved under the action of this scheme. In which case it follows immediately that any appropriate technique from conventional linear systems theory can be used to design $G_0^c(s)$ and, simultaneously, yield a candidate solution to the LPDP.

Stability (in the conventional sense) of the limit profile is a necessary condition for stability along the pass, for which interpreting theorem 3.3.7 closed-loop gives a set of necessary and sufficient conditions. Further, these conditions can be tested by applying either of the systematic procedures developed in section 4.2. Suppose also that theorem 3.3.7 holds closed-loop and consider the remaining specifications of the LPDP, i.e. those relating to the rate of convergence of $\{Y_k\}_{k \geq 1}$ to Y_∞ and the error $Y_k - Y_\infty, k \geq 0$, respectively. Then the only effective option at this stage is to undertake detailed simulation studies with the consequent prospect of a heavy computational load.

Note: In effect, the above approach to controller design examines two necessary conditions for stability along the pass, i.e. the limit profile and the derived conventional linear system. This is unavoidable, however, since stability testing based on the stability along the pass polynomial, $A_p(s,z)$, open or closed-loop is not a computationally feasible proposition.

As an alternative to using theorem 3.3.7, suppose that theorem 5.2.1, or theorem 5.2.2 in the unit memory case, holds closed-loop. Then use of these simulation-based tests produces, at no extra cost, unique computable information concerning the convergence rate of the $\{Y_k\}_{k \geq 1}$ and the error $Y_k - Y_\infty$, $k \geq 0$. This information results from interpreting theorem 5.3.1, or theorem 5.3.2 in the unit memory case, closed-loop under the assumption that the reference signal sequence is constant from pass to pass. Further, these facts lead immediately to the following candidate design procedure for solving the LPDP using memoryless dynamic unity-negative feedback control.

STEP 1: Use an appropriate technique from conventional linear systems theory to yield a candidate forward-path controller which satisfies the limit profile specifications. This step assumes that these specifications can, at least, be achieved to within 'acceptable bounds'. Further, the minimum specification in all cases is obviously stability in the conventional sense.

STEP 2: Interpret theorem 5.2.1, or theorem 5.2.2 in the unit memory case, in terms of the closed-loop system resulting from application of the controller designed at step 1. Then proceed to test this sufficient condition by applying the systematic procedure developed in section 5.2. On completion, the following two options exist.

STEP 3: If theorem 5.2.1, or theorem 5.2.2 in the unit memory case, holds the closed-loop system is stable along the pass. Then interpret theorem 5.3.1, or theorem 5.3.2 in the unit memory case, and proceed to step 5.

STEP 4: If theorem 5.2.1, or theorem 5.2.2 in the unit memory case does not hold, no definite conclusions can be drawn. The feasible options then are to

 (i)   test the necessary and sufficient conditions of theorem 3.3.7 interpreted closed-loop and, if they hold, proceed to simulation studies to assess the convergence rate and the error $Y_k - Y_\infty$, $k \geq 0$; or

 (ii) return to step 1 and re-design (if possible); or

 (iii) terminate

STEP 5: Decide if this design satisfies the specifications on the convergence rate and the error $Y_k - Y_\infty$, $k \geq 0$. If yes then stop, otherwise return to step 1 and re-design (if possible).

All steps in the above procedure are suitable for a CAD environment on the assumption that step 1 is completed using a compatible technique such as the characteristic locus or inverse Nyquist array. This is the subject of on-going research, see the cited reference for a complete update, which is developing the necessary infrastructure for inclusion in a user friendly interactive package, Coupled with this, research is proceeding in a number of general areas which, in effect, determine the effective operating range and efficiency in terms of, for example, computational cost of the above design procedure. These areas are not considered here and the cited reference again contains comprehensive details of progress to-date plus an update on some on-going current work. Instead, the following example is presented to illustrate the application, and potential, of this procedure.

Consider again the unit memory sub-class of (6.1) defined by the state-space model of (6.59). Then application of the above procedure to solve the LPDP for this case proceeds as follows.

STEP 1: Consider again the use of the proportional forward-path controller defined by (6.60). Then, in transfer-function matrix terms, it is easily shown that the closed-loop limit profile dynamics can be written as

$$(sI_m + \rho(I_m - \frac{B_0}{\rho}))Y_\infty(s) = \rho(I_m - \frac{A_0^{-1}A_1}{\rho})R_\infty(s) \tag{6.68}$$

which is stable in the conventional sense if, and only if, $\rho > \max_{1 \leq i \leq m} \text{Re}(\lambda_i)$ where λ_i is an eigenvalue of B_0. Consider also the case of $\rho \to +\infty$ ('high gain'). In which case $Y_\infty(t)$ is 'arbitrarily close' to the inverse Laplace transform of

$$Y_\infty(s) = \frac{\rho}{s+\rho} I_m R_\infty(s) \tag{6.69}$$

This is a totally non-interacting conventional linear system with zero steady-state error to a unit step applied at $t = 0$ in any channel.

STEP 3: In this case it is easily shown that the matrix $||L||_p$ of theorem 5.2.1 or 5.2.2 closed-loop, denoted $||L_c||_p$, is given by

$$||L_c||_p = \frac{1}{\rho} ||B_0||_p \tag{6.70}$$

This follows immediately since all entries in the step response matrix of the closed-loop associated conventional linear system are monotonic and sign definite, i.e. a special case of example 5.2.1. Hence theorem 5.2.1 holds $\forall \rho > r(||B_0||_p)$ and theorem 5.2.2 holds $\forall \rho > ||(||B_0||_p)||$. Interpreting theorem 5.3.1 now yields that the closed-loop output sequence $\{Y_k\}_{k \geq 1}$ approaches Y_∞ at a geometric rate governed by $\gamma \in (\frac{r(||B_0||_p)}{\rho}, 1)$, where this set is non-empty for any choice of $\rho > r(||B_0||_p)$, and interpreting (5.158) - (5.164) gives the error 'band' for each element of $Y_k - Y_\infty$, $k \geq 0$. Similarly, interpreting theorem 5.3.2 now yields that the closed-loop output sequence $\{Y_k\}_{k \geq 1}$ approaches Y_∞ at a geometric rate governed by $\frac{||(||B_0||_p)||}{\rho} < 1$ for suitable choice of ρ. Further, interpreting (5.148) - (5.150) gives the single error 'band' for $Y_k - Y_\infty$, $k \geq 0$.

STEP 5: Consider again the case of $\rho \to +\infty$ and, for example, the use of theorems 5.2.2 and 5.3.2. Then $||L_c|| = \frac{1}{\rho}||(||B_0||_p)|| \to 0$, i.e. the limit profile dynamics of (6.68) are reached to within arbitrary accuracy on the first pass in this case.

Note: The refinements of sections 5.2 and 5.3 in the form of, for example, theorem 5.2.5 and the results derived from lemma 5.3.1 are easily included in this design procedure.

In summary, therefore, this section has presented some initial results on solving the LPDP by memoryless output feedback control. The end product has been a systematic design procedure for solving this problem which is suitable for a CAD environment. Further, an in-depth treatment of its application to one sub-class has been given to highlight its potential. At this stage, it is not possible to fully assess the potential of this procedure, or the underlying control scheme, in terms of solving the LPDP. In particular, much work remains to be done in a number of areas for which the cited reference gives a comprehensive overview. One such area in the unit memory case is the development, and comparative studies of, an alternative procedure based, see also the discussion relating to (5.145) - (5.147), on combining all but the limit profile specifications into the single constraint $||L_c|| \leq b$, where $0 < b < 1$. Finally, note again that the memoryless schemes of section 6.1 have the simplest structure in implementation terms. Hence, as for stability along the pass, it is clear that the potential of these schemes should be fully evaluated before considering other possibilities.

6.4 The Repetitive Systems Disturbance Decoupling with Stability Problem

This section presents some initial results on the design of the schemes introduced in section 6.1 to solve the RSDDSP. In particular, the possibility of obtaining a 'geometric style' solution in the spirit of that for the well known conventional linear systems problem is briefly explored. Further, the memoryless dynamic unity-negative feedback controller with proportional repetitive minor loop compensation introduced in section 6.1 is used to solve this problem in one special case. Finally, some possible future research topics are noted.

As a preliminary to the discussion which immediately follows, it is instructive to briefly review the well-known disturbance decoupling with stability problem for conventional linear systems. Consider, therefore, the system of (6.71) below where $q(t)$ represents a disturbance which is assumed not to be directly measurable by the controller

$$\dot{X}(t) = AX(t) + BU(t) + Dq(t)$$
$$Y(t) = CX(t)$$
$$X(t) \in R^n, \ Y(t) \in R^m, \ U(t) \in R^\ell, \ q(t) \in R^v \qquad (6.71)$$

Further, suppose that the linear state feedback law $U(t) = FX(t)$ is applied to (6.71). Then the disturbance decoupling problem for the resulting closed-loop system is to find a suitable F such that $q(t)$ has no influence on the controlled output $Y(t)$. Equivalently, this closed-loop system is said to be disturbance decoupled relative to the pair $Y(t)$, $q(t)$ if, for each initial condition $X(0) \in R^n$, the output $Y(t)$, $t \geq 0$, is the same for all $q(t) \in R^v$.

The above problem has been the subject of much research effort. One element of which has been to use such geometric concepts as (A,B)-invariant subspaces to develop necessary and sufficient conditions for the existence of a solution which are, for example, given in the cited reference. Note, however, that these conditions do not guarantee closed-loop stability, i.e. that all eigenvalues of A+BF have strictly negative real parts, which is obviously essential for applications. This has led to the so-called disturbance decoupling with stability problem for which the cited reference also gives necessary and sufficient conditions for the existence of a solution.

Return now to the repetitive systems case and, since the following discussion generalises in a natural manner, consider the special case of a unit memory process. Further, interpret $Y_k(t)$, $0 \leq t \leq \alpha$, $k \geq 0$, as a disturbance which is not directly measured by the controller on pass $k + 1$. Suppose also that the current pass linear state feedback law

$$U_{k+1}(t) = F \, X_{k+1}^P(t), \quad 0 \leq t \leq \alpha, \quad k \geq 0 \qquad (6.72)$$

is applied. Then a clear structural similarity exists with the conventional linear systems problem for (6.71). To see this, consider again the first requirement of the RSDDSP and interpret it in terms of the closed-loop system. In which case it follows immediately that this requirement holds relative to the pair $Y_{k-1}(t)$, $Y_k(t)$, $0 \leq t \leq \alpha$, $k \geq k^* \geq 1$, if, for each initial condition $X_k^P(0) = d_k \in R^{n_1}$, the output $Y_k(t)$ is the same for all $Y_{k-1}(t) \in R^m$. Hence repetitive systems disturbance decoupling simply means that the contribution of the previous pass profile to the current one is zero, $0 \leq t \leq \alpha$, $k \geq k^* \geq 1$. The second requirement of the RSDDSP for this case requires, as a basic minimum, that F be selected such that all eigenvalues of $A^P + B^P F$ have strictly negative real parts. These facts now lead immediately to the conclusion that the RSDDSP in this case is structuraly similar to its well researched conventional linear systems counterpart. Further, it can be conjectured that a solution to this RSDDSP can be developed using geometric concepts such as (A,B)-invariant subspaces with the consequent possibility of a natural generalisation to the non-unit memory case. This general area is the subject of on-going research for which the cited reference gives a comprehensive treatment of the considerable progress to-date.

On the assumption that a solution exists, implementation of a current pass state feedback solution to the RSDDSP would encounter the same potential difficulties as other uses of this law. In which case one option is to use output feedback based schemes. The analysis below uses the memoryless dynamic unity-negative feedback controller with proportional repetitive minor loop compensation introduced in section 6.1 to solve the RSDDSP in one special case. This analysis also represents the first output on the design of these minor loop schemes or, more generally, controllers with memory.

Return, therefore, to the closed-loop state-space model of (6.50) which results from application of the control law of (6.46)-(6.47) to (6.1). Consider also the special case when $D_j^P = 0$, $1 \leq j \leq M$, $n_1 = m = \ell$ and $|B^P| \neq 0$; a not uncommon situation in industrial examples such as bench mining systems. Further, select the controller matrices K_j as

$$K_j = (B^P)^{-1} B^P_{j-1}, \ 1 \leq j \leq M \tag{6.73}$$

Then $B^P_{j-1} = 0$, $1 \leq j \leq M$, in (6.49) and hence in (6.50) the pass profile $Y_k(t)$, $0 \leq t \leq \alpha$, is independent of the pass profiles $Y_{k-j}(t)$, $1 \leq j \leq M$, for all passes $k \geq 1$. Equivalently, repetitive systems disturbance decoupling is achieved in this case with the optimum choice of $k^* = 1$.

Suppose, therefore, that (6.73) holds. Then it is easily shown that the closed-loop limit profile is described in transfer-function matrix terms by

$$Y_\infty(s) = (I_m + G^P_0(s) G^C_0(s))^{-1} G^P_0(s) G^C_0(s) R_\infty(s) \tag{6.74}$$

where $G^P_0(s)$ and $G^C_0(s)$ are defined by (6.19) and (6.22) respectively. Equivalently, the limit profile is described by the derived conventional linear system. Hence the design exercise can be completed by using an appropriate technique to design $G^C_0(s)$ to meet the required specifications.

In summary, therefore, this section has presented some preliminary work on solving the RSDDSP by use of the schemes introduced in section 6.1. At best, these demonstrate the potential of these schemes in terms of this problem and it is not possible at this stage to fully assess them in this context. This can only take place after much further work has been undertaken in a number of areas which are detailed in the cited reference.

Notes and References

The three control policies of section 6.1 are from Rogers and Owens (1990 i,j,k) respectively. The optimal control problem referred to is from Willson, Collins and Owens (1982) and Rogers and Owens (1990i) also details other candidate control policies. All control schemes detailed or referred to in this section are from Rogers and Owens (1990ℓ). Owens (1978) and Wonham (1974), together with the relevant references therein, are two of numerous possible sources for the cited results from conventional linear systems theory. As noted previously, Rogers and Owens (1990b) details the work to date on poles and Smyth, Rogers and Owens (1990a) discusses some implementation aspects of the control schemes introduced to - date. Rogers and Owens (1990m) summarises the corresponding analysis of this section for the discrete case.

Section 6.2 is based on Rogers and Owens (1990i) and Smyth, Rogers and Owens (1990b,c). Use has also been made of results from Rogers and Owens (1988a, 1989d) and Smyth (1991).

Section 6.3 is based on Rogers and Owens (1990i,j) and Smyth (1991). Finally, section 6.4 is from Rogers and Owens (1990k) and has also made use of results from Rogers and Owens (1988b).

CHAPTER 7

CONCLUSIONS AND FURTHER WORK

Using previous work as a basis, a rigorous stability theory for repetitive processes with linear dynamics and a constant pass length has been presented. This has been formulated in terms of a general abstract representation which, in effect, regards the output on any pass as a point in a Banach space. Further, this model includes as special cases all unit and non-unit memory repetitive processes with linear dynamics and a constant pass length. Hence an obvious way to develop a rigorous stability theory is to formulate this in terms of the abstract model and then interpret the resulting conditions in terms of the particular example under consideration.

The resulting stability theory consists of two distinct concepts termed asymptotic stability and stability along the pass respectively. Further, asymptotic stability is a necessary condition for stability along the pass which is required in all practical applications. To provide a basic explanation of this fact, recall that the essential unique control problem for these processes is the possible presence in the output sequence of oscillations which increase in amplitude from pass to pass. Then, in effect, asymptotic stability guarantees the existence of the limit profile as a function of the (finite) pass length and stability along the pass is independent of this parameter. This, in turn means that asymptotic stability alone would permit exponential growth terms in the dynamics along a pass - an obviously totally undesirable feature.

Necessary and sufficient conditions for stability are expressed in terms of the spectral radius and resolvent of the linear operator associated with the abstract representation. Hence application of this theory to a particular example requires the interpretation of these results in terms of the parameters of its representation or model. No general rules exist for this task and severe difficulties could arise if the underlying Banach space or the linear operator have a complex structure. A significant number of industrially relevant special cases can, however, be dealt with in a relatively simple manner. This has been demonstrated here by a detailed consideration of the long-wall coal cutting example and differential and discrete non-unit memory linear repetitive processes. In these latter two cases, the resulting conditions are expressed in terms of the matrices of the corresponding state-space descriptions.

The basic difficulty with these conditions is that testing one of them is not computationally feasible, where this is a common feature of the results to-date for a number of other cases. Further, given the pivotal role of stability, the development of computationally feasible stability tests is an obvious starting point for any further control related analysis of a given case. Consequently a substantial part of the work reported in this monograph has been the development of

computationally feasible stability tests for differential and discrete non-unit
memory linear repetitive processes.

Previously reported work has established strong structural links between a
number of sub-classes and other well researched classes of dynamic systems. In the
case of differential and discrete processes, such links have been established with
the following two classes of linear dynamic systems.

(i)   Standard or, within the repetitive systems framework, conventional linear
      systems described by the well known state-space model or transfer-function
      matrix. Basically, the repetitive systems state-space model reduces to
      its conventional linear systems counterpart under well defined conditions.

(ii) 2D linear systems described by the Roesser state-space model. Basically,
      stability along the pass in the discrete unit memory case is equivalent to
      BIBO stability of the Roesser model.

All of the new results presented in this monograph have, in effect, been
developed by appropriately exploiting these links or, in the differential case,
results from the stability theory of certain classes of delay differential systems.

As a result of this approach, three distinct types of computationally feasible
stability tests have been developed which can be classified under the following
general headings.

(i)   Graphical, or eigenvalue, based tests.

(ii)   Algebraic, or root clustering, based tests.

(iii) Simulation-based tests.

Further, (i) and (ii) share a common basis in that they are both based on a
reformulation of the original state-space conditions in terms of the appropriate 2D
transfer-function matrix. These new conditions have then been used to develop two
systematic test procedures for each case which test them in a particular order with
termination if the one just tested does not hold. Note also that in all cases the
computationally most intensive condition is the last to be tested (if required).

The systematic procedure for each case based on (i) above uses, in effect,
'Nyquist like' tests from the stability theory of differential and discrete
conventional linear systems as appropriate. Hence these procedures are suitable for
inclusion in a CAD package. As an alternative, the systematic procedures based on
(ii) above make appropriate use of well known conventional linear systems root
clustering based stability tests. For example, the procedure for the discrete case
uses the Jury/Marden table and the one for the differential case uses the Routh
array and its modified version for real even order polynomials. These procedures
are not suitable for inclusion in a CAD package and have their major remit in low
order synthesis problems where some, or all, of the elements of the matrices of the
example under consideration are design parameters.

Previous work has shown that BIBO stability of 2D linear systems described by
the Roesser model is equivalent to stability along the pass of discrete unit memory
linear repetitive processes. Consequently all of the known stability tests for

these systems can be applied to the repetitive systems problem. In this work, however, particular attention has been directed towards the use of Lyapunov equations. This has shown that, unlike the conventional linear systems case, two essentially different approaches are possible. One of these is based on a 2D equation with constant coefficients and the other uses a 1D equation with coefficients which are functions of a complex variable.

Either of these Lyapunov equations could be used to form the basis of a systematic test procedure to serve as an alternative to those based on (i) and (ii) above. Note, however, that the constant coefficient version is, in general, sufficient but not necessary and this fact clearly reduces its effectiveness given the alternatives which test necessary and sufficient conditions. Detailed in depth comparative studies of all of these procedures for the discrete unit memory case has, however, not been considered in the absence to date of results from applying them to suitably defined benchmark problems.

The use of the constant coefficient Lyapunov equation in developing physically meaningful stability margins for the discrete unit memory case has been considered. This initial work has, in effect, been based on extending some results from 2D linear systems. Further, there are two (interrelated) areas to which future research effort could profitably be directed. These are further development of the basic computational algorithm for increased efficiency and in depth work to establish the correlation with system performance. Progress in both of these areas will also obviously serve to strengthen the already known links between these two, apparently distinct, areas. Note also that it is by no means clear how, if at all, the procedures based on (i) and (ii) above can be exploited in terms of stability margins, except as a result of extensive simulation studies following on from the basic testing.

In the differential unit memory case, it has been shown that elements of the stability theory of delay differential systems can be used to produce systematic test procedures as alternatives to those based on (i) and (ii) above. This arises from a proof of a previously conjectured result that stability along the pass is equivalent to pointwise asymptotic stability when this example is interpreted as a delay differential system. Hence all of the known stability tests for delay differential systems can be applied to this example but here, as in the discrete case, particular attention has been directed towards the use of Lyapunov equations. The resulting analysis again yielding two essentially different approaches, based on a 2D equation with constant coefficients and a 1D equation with coefficients which are functions of a complex variable respectively.

As in the discrete case, either of these Lyapunov equations could be used to form the basis of an alternative systematic test procedure to those based on (i) and (ii) above. Again, however, the constant coefficient version is, in general, sufficient but not necessary and this fact clearly reduces its effectiveness given the alternatives which test necessary and sufficient conditions. Detailed comparisons of all the available procedures for this case would, however, require results from applying them to suitably defined benchmark problems. Hence this topic has not been considered further in this work.

Using an analogous approach to the discrete case, the constant coefficient version has been used to develop some initial results on physically meaningful stability margins. The results presented are, in effect, appropriate extensions of some of those currently available for delay differential systems. Further, two (interrelated) areas have been identified to which future research effort could profitably be directed. These are the development of a computational algorithm (which does not yet exist for delay differential systems) and in depth work to establish the correlation with systems performance. Progress in these two areas will again also serve to strengthen the links between these two, apparently distinct, areas. Finally, note that, as in the discrete case, it is by no means clear how, if at all, the procedures based on (i) and (ii) above can be exploited in terms of stability margins, except as a result of extensive simulation studies following on from the basic testing.

In terms of controlling these processes, it is known that computable information concerning the following factors is of at least equal importance to stability margins.

(i) The rate of approach of the output sequence to the limit profile.

(ii) Bounds on the performance along any pass.

Suppose also that the simulation-based tests listed under (iii) earlier in this chapter are used. Then these tests produce, at no extra cost, computable information on (i) and (ii) above. This information is unique to these tests, with the only possible alternative being to inspect the results of simulation studies and hence the prospect of a heavy computational load. Further, this feature serves to offset the fact that these tests are sufficient but not necessary.

These simulation-based tests have been developed from suitably well behaved plant step response data which, in this work, is assumed to be available or can be obtained by simulation studies. In particular, it has been assumed in this initial study that the parameters in the sub-processes which define this step response data (the derived and conventional linear systems) are known exactly. By analogy with the use of such information in conventional linear systems (one form of robust control analysis) it can be conjectured that this assumption can be relaxed in a relatively straightforward manner. Further, some initial highly promising results on extending these tests to processes with interpass smoothing have been presented. These are the first reported output on the analysis of such cases and it is by no means clear at this stage how, if at all, tests based on other approaches can be effectively used in this context. Hence it is clear that the simulation-based approach to the control related analysis of certain sub-classes of linear repetitive processes should, after appropriate development, be a powerful and flexible technique.

Return now to the basic problem of testing a differential or discrete non unit memory linear repetitive process for stability. Then the work reported here has produced a range of computationally feasible stability tests grouped under three general headings. Note, however, that much work remains to be done which can be broadly classified as follows.

(a) Further development of the specific areas detailed as appropriate in the main text plus work on extending their effective operating range. For example, further work on applying the simulation-based tests to processes with interpass smoothing effects should yield rapid progress.

(b) Further work aimed at achieving maximum efficiency coupled, in the case of the eigenvalue and simulation-based tests, with the development of a software infrastructure to form the basis of a comprehensive computer aided analysis/design package. This should use currently available software as a basis and include the development of a suitable user interface facility.

(c) The development of 'easy to use' stability margins and/or performance indicators to assist in the formulation and solution of practically relevant control policies. This area should initially proceed from the highly promising work reported in chapters 4 and 5 for this general area.

(d) The development of a comprehensive 'systems theoretic' interpretation of stability and related matters. One possible use of such a theory would be to provide indicators of control difficulties in the spirit of Wonham's controllability result for pole assignment in conventional linear systems. Specific areas for initial work could include (i) the precise implications of the controllability and observability conditions of the necessary and sufficient stability results of chapter 3, (ii) the precise roles (if any) of appropriately defined poles and zeros where one candidate definition of the former in the unit memory cases has already been proposed, and (iv) the derivation of stability conditions expressed in terms of the 2D transfer-function matrices as an entity instead of as in this work where only its constituent elements have been used (in particular, the transfer-function matrices of the derived and associated conventional linear systems).

The success in developing basic computationally feasible stability tests has led to some initial work on controller design. In particular, three control policies have been formulated from practical considerations and feedback control schemes which use either state or output information have been developed. Further, some candidate design algorithms have been developed together with some relevant systems theoretic properties.

At a general level, the schemes developed in this work have demonstrated the potential power of appropriately specified feedback control schemes in regulating the behaviour of differential and discrete non-unit memory linear repetitive processes. Further, an attractive feature of these schemes from an applications standpoint is their relative simplicity and hence the possibility of implementation without recourse to special purpose hardware/software. This is particularly true for the memoryless cases since they have the simplest structure in terms of the information to be logged and/or stored in order to actuate the controller. Hence it is strongly recommended that the potential of these schemes should be fully evaluated before recourse to others with a more complex structure. As a starting point, the specific areas detailed as appropriate in chapter 6 should be addressed.

In conclusion, therefore, substantial progress towards the development of rigorous stability and control theories for differential and discrete non-unit memory linear repetitive processes has been made based, essentially, on an abstract representation of the general linear dynamics constant pass length case. This strongly suggests that a similar approach to other general cases should prove equally successful, particularly if experience gained in the course of the work reported here can be exploited. One obvious area which should benefit considerably from such an approach is that of a constant pass length and certain classes of nonlinear dynamics.

REFERENCES

Agathoklis, P. (1988). "Lower Bounds for the Stability Margin of Discrete Two-Dimensional Systems Based on the Two-Dimensional Lyapunov Equation", IEEE Trans. Circuits and Systems, Vol. CAS 35, No. 6, pp.745-749.

Agathoklis, P. and Foda, S. (1989a). "Stability and the Matrix Lyapunov Equation for Delay Differential Systems", Int. J. Control, Vol. 49, No. 2, pp.417-432.

Agathoklis, P. and Foda, S. (1989b). "Lower Bounds for the Stability Margin of Delay Differential Systems", Proc ICASS 89, pp.549-552.

Agathoklis, P., Jury, E.I. and Mansour, M. (1989). "The Discrete-Time Strictly Bounded-Real Lemma and the Computation of Positive Definite Solutions to the 2-D Lyapunov Equation", IEEE Trans. Circuits and Systems, Vol. CAS 36, No. 6, pp.830-837.

Agathoklis, P., Jury, E.I. and Mansour, M. (1990). "An Algebraic Test for Internal Stability of 2-D Discrete Systems", in Kaashoek, M.A. etal Eds 'Realisation and Modelling in System Theory', Birkhauser: Boston, pp.303-310.

Anderson, B.D.O. and Vongpanithlerd, S. (1973). "Network Analysis and Synthesis, A Modern Systems Theory Approach", Prentice-Hall: Englewood Cliffs N.J.

Anderson, B.D.O., Agathoklis, P., Jury, E.I. and Mansour, M. (1986). "Stability and the Matrix Lyapunov Equation for Discrete 2-Dimensional Systems", IEEE Trans. Circuits and Systems, Vol. CAS 33, No. 3, pp.261-266.

Boland, F.M. and Owens, D.H. (1980). "Linear Multipass Processes - A Two-Dimensional Interpretation", Proc. IEE, 127, (5), pp.189-193.

Edwards, J.B. (1974). "Stability Problems in the Control of Multipass Processes", Proc. IEE, 121, (11), pp.1425-1431.

Edwards, J.B. and Owens, D.H. (1982). "Analysis and Control of Multipass Processes", Wiley Research Studies Press: Chichester.

Fadali, M.S. and Gnanasekaran, R. (1989). "Normal Matrices and their Stability Properties: Application to 2-D System Stabilisation", IEEE Trans. Circuits and Systems, Vol. CAS 36, No. 6, pp.873-875.

Gantmacher, F.R. (1959). "The Theory of Matrices Vols I and II", Chelsea: New York.

Gu, G. and Lee, E.B. (1989). "Stability Testing of Delay Differential Systems", Automatica, Vol. 25, (2), pp.777-780.

Hale, J.K. (1977). "Theory of Functional Differential Equations", Springer-Verlag: New York.

Huang, T.S. (1972). "Stability of Two-dimensional and Recursive Filters", IEEE Trans. Audio and Electroacoustics, Vol. AU-20, (2), pp.158-163.

Jury, E.I. (1974). "Inners and the Stability of Dynamic Systems", Wiley: New York.

Kamen, E.W. (1982). "Linear Systems with Commensurate Time Delays: Stability and Stabilisation Independent of Delay", IEEE Trans. Auto Control, Vol. 27, No. 2, pp.367-375.

Lancaster, P. and Tismenetsky, M. (1985). "The Theory of Matrices", Academic Press.

Lu, W.S. and Lee, E.B. (1985). "Stability Analysis for Two-Dimensional Filters via a Lyapunov Approach", IEEE Trans. Circuits and Systems, Vol. CAS 32, No. 11, pp.61-68.

Owens, D.H. (1977). "Stability of Linear Multipass Processes", Proc. IEE, 124, (11), pp.1079-1082.

Owens, D.H. (1978). "Feedback and Multivariable Theory", Peter Peregrinus: London.

Owens, D.H. and Chotai, A. (1983). "Robust Controller Design for Linear Dynamic Systems Using Approximate Models", Proc. IEE, 130, (2), pp.45-56.

Piekarski, M.S. (1977). "Algebraic Characterisation of Matrices whose Multivariable Characteristic Polynomial is Hurwitzian", Proc. Int. Symp. Operator Theory, Lublock TX, pp.121-126.

Postlethwaite, I. and MacFarlane, A.G.J. (1979). "A Complex Variable Approach to the Analysis of Linear Multivariable Systems", Springer Verlag Lecture Notes in Control and Information Sciences, Vol. 12: Berlin.

Roesser, R.P. (1975). "A Discrete State-Space Model for Linear Image Processing", IEEE Trans. Auto-Control, Vol. AC-20, No. 1, pp.1-10.

Rogers, E. (1987). "Feedback and Stability Theory for Linear Multipass Processes", A series of Research Reports, The Queen's University of Belfast.

Rogers, E. and Owens, D.H. (1988a). "Stability and State Feedback Control of Differential Unit Memory Linear Multipass Processes", Proc. 1988 ACC, Vol. 1, pp.51-52.

Rogers, E. and Owens, D.H. (1988b). "Controller Design for Industrial Multipass Processes", Proc 3rd European Conference for Mathematics in Industry, Glasgow, pp.495-502.

Rogers, E. and Owens, D.H. (1989a). "2D Transfer-Functions and Stability Tests for Differential Non-Unit Memory Linear Multipass Processes". Int. J. Control, Vol. 50, No. 2, pp.651-666.

Rogers, E. and Owens, D.H. (1989b). "Axis Positivity and the Stability of Linear Multipass Processes", "Linear Algebra and its Applications", Vol's. 122/123/124, pp.779-796.

Rogers, E. and Owens, D.H. (1989c). "Stability Analysis for Discrete Linear Multipass Processes with Non-Unit Memory", IMA Journal of Mathematical Control and Information, Vol. 6, No. 4, pp,399-409.

Rogers, E. and Owens, D.H. (1989d). "Output Feedback Control of Linear Multipass Processes", Proc. 1989 ACC, Vol. 1, pp318-319.

Rogers, E. and Owens, D.H. (1990a). "2D Transfer-Functions and Stability Tests for Discrete Linear Multipass Processes", in Kaashoek, M.A. etal Eds, "Realisation and Modelling in System Theory", Birkhauser: Boston, pp.351-356.

Rogers, E. and Owens, D.H. (1990b). "Poles and Related Matters for Differential and Discrete Linear Repetitive Processes". Research Report No. DC15, University of Strathclyde, Division of Dynamics and Control.

Rogers, E. and Owens, D.H. (1990c). "Stability of Linear Repetitive Processes: a Lyapunov Approach". Research Report No. DC16, University of Strathclyde, Division of Dynamics and Control.

Rogers, E. and Owens, D.H. (1990d). "Simulation-based Stability Tests for Differential Non-unit Memory Linear Multipass Processes", Submitted to IEEE Trans. Auto Control.

Rogers, E. and Owens, D.H. (1990e). "Simulation-based Stability Tests for Differential Unit Memory Linear Multipass Processes", Int. J. Control, Vol. 51, No. 5, pp.1051-1066.

Rogers, E. and Owens, D.H. (1990f). "Improved Stability Tests and Performance Bounds for Differential Linear Repetitive Processes", Research Report No. DC17, University of Strathclyde, Division of Dynamics and Control.

Rogers, E. and Owens, D.H. (1990g). "A Comprehensive Stability Analysis for Discrete Linear Repetitive Processes", Submitted to Multi-Dimensional Systems and Signal Processing.

Rogers, E. and Owens, D.H. (1990h). "Stability Analysis for Repetitive Processes with Interpass Smoothing", Research Report No. DC18, University of Strathclyde, Division of Dynamics and Control.

Rogers, E. and Owens, D.H. (1990i). "Stability based Control Policies for Differential Linear Repetitive Processes", Research Report No. DC19, University of Strathclyde, Division of Dynamics and Control.

Rogers, E. and Owens, D.H. (1990j). "The Limit Profile Design Problem for Differential Linear Repetitive Processes", Research Report No. DC20, University of Strathclyde, Division of Dynamics and Control.

Rogers, E. and Owens, D.H. (1990k). "The Repetitive Systems Disturbance Decoupling with Stability Problem", Research Report No. DC21, University of Strathclyde, Division of Dynamics and Control.

Rogers, E. and Owens, D.H. (1990ℓ). "Control Schemes for Differential Repetitive Processes", Research Report No. DC22, University of Strathclyde, Division of Dynamics and Control.

Rogers, E. and Owens, D.H. (1990m). "Control Policies and Schemes for Discrete Linear Repetitive Processes", Research Report No. DC23, University of Strathclyde, Division of Dynamics and Control.

Shanks, J.L., Treitel, S. and Justice, J.H. (1972). "Stability and Synthesis of Two-Dimensional and Recursive Filters", IEEE Trans. Audio and Electroacoustics, Vol. AU20, No. 2, pp.115-128.

Siljak, D.D. (1971). "New Algebraic Criteria for Positive Realness". Journal of the Franklin Institute, Vol. 291, pp.109-120.

Siljak, D.D. (1973). "Algebraic Criteria for Positive Realness Relative to the Unit Circle", Journal of the Franklin Institute, Vol. 295, pp.469-476.

Siljak, D.D. (1975). "Stability Criteria for Two Variable Polynomials", IEEE Trans. Circuits and Systems, Vol. CAS22, No. 3, pp.185-189.

Smyth, K.J. (1991). "Computer Aided Analysis of Linear Repetitive Processes", Ph.D. Thesis, University of Strathclyde.

Smyth, K.J., Rogers, E. and Owens, D.H. (1990a). "Some Issues Relating to the Implementation of Control Schemes for Linear Repetitive Processes", Research Report No. DC 24, University of Strathclyde, Division of Dynamics and Control.

Smyth, K.J., Rogers, E. and Owens, D.H. (1990b). "Multivariable First Order Lag Models for the Control of Linear Repetitive Processes", Research Report No. DC25, University of Strathclyde, Division of Dynamics and Control.

Smyth, K.J., Rogers, E. and Owens, D.H. (1990c). "Simulation-based Tests in Controller Design for Stability along the Pass of Differential Linear Repetitive Processes", Research Report No. DC 26, University of Strathclyde, Division of Dynamics and Control.

Strintzis, M.G. (1977). "Tests of Stability of Multi dimensional Filters", IEEE Trans. Circuits and Systems, Vol. CAS24, No. 8, pp.432-437.

Taylor, A.E. (1958). "Introduction to Functional Analysis", Wiley: New York.

Willems, J.L. (1970). "Stability Theory of Dynamical Systems", Nelson: London.

Willson, I.W., Collins, W.D. and Owens, D.H. (1982). "Optimal Control of Linear Differential Multipass Processes", Research Report No. 173, Universiy of Sheffield, Department of Control Engineering.

Wonham, W.M. (1974). "Linear Multivariable Control: A Geometric Approach", Springer Verlag: New York.

Lecture Notes in Control and Information Sciences

Edited by M. Thoma and A. Wyner

Lecture Notes in Control and Information Sciences

Edited by M. Thoma and A. Wyner

Lecture Notes in Control and Information Sciences

Edited by M. Thoma and A. Wyner

Vol. 174: A.J.M. Beulens, H.-J. Sebastian (Eds.)
Optimization-Based Computer-Aided
Modelling and Design
Proceedings of the First Working Conference
of the IFIP TC 7.6 Working Group,
The Hague, The Netherlands, 1991
VIII, 270 pages, 1992

Vol. 175: E. Rogers, D.H. Owens
Stability Analysis for Linear Repetitive Processes
VII, 197 pages, 1992